J. H. J. Peet

B.Sc., M.Sc., Ph.D., C.Chem

Further studies in chemistry

A chemistry textbook for TEC level 3

Longman London and New York

Longman Group Limited London

Associated companies, branches and representatives throughout the world

Published in the United States of America by Longman Inc., New York

© Longman Group Limited 1980

First published 1980

British Library Cataloguing in Publication Data

Peet, J H J
    Further studies in chemistry. — (Longman
    technician series : mathematics and sciences).
    1. Chemistry
    I. Title
    540'.2'46 QD31.2 79—40207

    ISBN 0—582—41172—6

Printed in England by McCorquodale (Newton) Ltd., Newton-le-Willows, Lancashire.

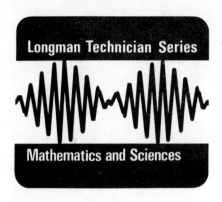

Longman Technician Series

Mathematics and Sciences

Sector Editor:

D. R. Browning, B.Sc., F.R.I.C., A.R.T.C.S.
Principal Lecturer and Head of Chemistry, Bristol Polytechnic

Books already published in this sector of the series:

Technician mathematics Level 1   **J. O. Bird** *and* **A. J. C. May**
Technician mathematics Level 2   **J. O. Bird** *and* **A. J. C. May**
Technician mathematics Level 3   **J. O. Bird** *and* **A. J. C. May**
Mathematics for science technicians Level 2   **J. O. Bird** *and* **A. J. C. May**
Mathematics for electrical and telecommunications technicians Level 2
       J. O. Bird *and* **A. J. C. May**
Mathematics for electrical technicians Level 3   **J. O. Bird** *and* **A. J. C. May**
Calculus for technicians   **J. O. Bird** *and* **A. J. C. May**
Physical sciences Level 1   **D. R. Browning** *and* **I. McKenzie Smith**
Engineering science for technicians Level 1   **D. R. Browning** *and*
       I. McKenzie Smith
Safety science for technicians   **W. J. Hackett** *and* **G. P. Robbins**
Fundamentals of chemistry   **J. H. J. Peet**
Engineering science in SI units, 2nd edition   **E. Hughes** *and* **C. Hughes**
Mathematics for scientific and technical students   **H. G. Davies** *and*
       G. A. Hicks
Mathematical formulae for TEC courses   **J. O. Bird** and **A. J. C. May**

Contents

Preface

This volume is written as a sequel to the author's *Fundamentals of Chemistry* (Longman, 1978). It is designed to cover the theoretical chemistry necessary for the Technician Education Council level III units in Chemistry, specifically Chemistry III and Laboratory Techniques (physical science) III. The book also covers the work required for Analytical Chemistry II.

A selection of tested experiments has been included in each chapter. These experiments have been chosen to illustrate the concepts developed in the main text. The SI convention of units and the IUPAC system of nomenclature are followed. Physical data quoted are generally taken from the *Nuffield Advanced Science Book of Data* (Longman, 1975).

Each chapter concludes with a list of references for further study, as well as suggested films and questions.

The author thanks Mr David R. Browning, editor of this Longman Technician Series (Mathematics and Sciences), for his valued guidance.

J. H. J. Peet

Acknowledgements

We are grateful to the following for permission to reproduce copyright material:
American Chemical Society for fig. by B. G. Harvey *et al* p. 6229 *American Chemical Society Journal* Vol. 76 and fig. by L. J. Gruggenberger and R. E. Rundle p. 5344 *American Chemical Society Journal* Vol. 86; City and Guilds of London Institute for data from question 6, paper 315−2−05 *Chemical Technician's Advanced Certificate Examination 1972*; Pergamon Press for fig. 6.5 p. 168 from *Analytical Applications of Ion Exchangers* by J. Inczédy, translated from Hungarian by A. Pall (Pergamon 1966); Taylor and Francis Ltd and J. R. Majer for fig. 1.8(a) based on fig. 1.17 p. 24 by J. R. Majer and M. Berry from *The Mass Spectrometer.*

Part 1

Physical chemistry

Chapter 1

Structure

Revision reading

The background to this chapter is covered in *Fundamentals of Chemistry*, Chapters 1 to 4.

Electronic structure

Electrons occupy the region around the nucleus of an atom. As a result of their small size, it is not possible to view the electrons directly and so their positions and movement have to be deduced from their effects. Some of the behaviour of the electrons is more easily interpreted in terms of wave motion than by using a simple particle explanation. This led de Broglie to postulate that a particle in motion had wave properties (Fig. 1.1) which can be represented by the formula

$$\lambda = \frac{h}{m \cdot v}$$

where λ = wavelength(m), m = mass(kg), v = velocity(m s^{-1}) and h is a constant (**Planck's constant**) with a value of 6.62×10^{-34} J s (joule-second).

Consider an electron of mass 9.109×10^{-31} kg and velocity 2.998×10^8 m s^{-1}. According to the de Broglie formula, it must have a wavelength of 2.424×10^{-12} m.

Because of the low mass of the electron, there is a significant uncertainty entailed in defining both the position and the momentum with high precision.

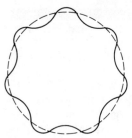

Fig. 1.1 The wave properties of a particle in motion

The reason for this is that, in order to examine the electron, electromagnetic radiation must interact with it. The energy of this radiation gives it a substantial 'kick', deflecting it from its original position. Let us consider two situations by way of illustration. First of all, if visible radiation is directed on to an electron, its momentum is tranferred to the particle. Using visible light radiation of energy 4×10^{-19} J, its momentum will be given by

momentum $= m \cdot v = $ energy$/\frac{1}{2}v$

where the energy is $\frac{1}{2}m \cdot v^2$. So, the momentum $= \dfrac{4 \times 10^{-19}}{0.5 \times 3 \times 10^8}$ kg m s^{-1},

i.e. 2.7×10^{-27} kg m s^{-1}. If all the momentum is transferred to the electron, this will move with a velocity given by

velocity $= \dfrac{\text{momentum}}{\text{mass}} = \dfrac{mv}{m}$

$= (2.7 \times 10^{-27})/(9 \times 10^{-31})$ m s^{-1}

$= 3 \times 10^3$ m s^{-1}

So, it is deflected from its original position at a rate of 3 km per second. So, there is not much chance of measuring its position very precisely!

A second example shows why the mass of the particle is important. The same interaction with a 0.5 kg ball gives it a velocity of

$\dfrac{2.7 \times 10^{-27}}{0.5} = 5 \times 10^{-27}$ m s^{-1}

That is not likely to create much difficulty in measuring the position precisely! Heisenberg presented a principle based on this work in which he states that 'it is impossible to determine precisely both the position and the momentum of a particle such as the electron'. The more precisely its velocity is known (and so its momentum), the less precisely its position can be determined.

In order to overcome these problems and to give us some understanding of the electron's dynamic behaviour, Schrödinger developed a description of the energy of the electron. The equations can be found in advanced textbooks, but they are too difficult to consider here; we are considering only the results. His equations are based on the following premises:

4

(a) electrons possess wave characteristics;
(b) the positions of the electrons cannot be described with a high degree of precision if the energy is to be quoted accurately;
(c) the electrons can only absorb specific quantities of energy (as described by the quantum theory).

Using these assumptions, Schrödinger developed a series of mathematical expressions from which it is possible to describe the energies and most probable positions of the electrons in an atom. Four terms, **quantum numbers**, are used to describe these electrons.

(a) The principal quantum number

The first of these terms is known as the principal quantum number and is given the symbol n. Numerically, n can have integer values 1, 2, 3 etc. In general, a lower value of n is possessed by an electron which is more strongly attracted to the nucleus than by an electron which is less strongly held by it. The values of n are also given the letters K, L, M . . .

(b) The azimuthal quantum number

The azimuthal quantum number, also known as the subsidiary quantum number, is given the symbol l. Though it can be described by numerical values (0, 1, 2 . . .), it is more normally designated by a series of letters: s ($l = 0$), p ($l = 1$), d ($l = 2$), f ($l = 3$). The letters chosen were derived from spectroscopic terms. The number of values of l for each principal quantum number is given by n. For example, for $n = 1$ there is one value of l; when $n = 2$, there are two values of l; three values of l occur when $n = 3$. Table 1.1 lists the acceptable combinations of n and l for the first four values of the

Table 1.1   Combinations of the principal and azimuthal quantum numbers for $n = 1$ to 4

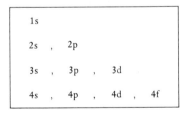

1s			
2s ,	2p		
3s ,	3p ,	3d	
4s ,	4p ,	4d ,	4f

principal quantum number. The lower values of l correspond to electrons with the largest attractive forces between themselves and the nucleus.

(c) The magnetic quantum number

It is found that electrons are affected differently by applied magnetic fields. For example, electrons which can be described by the term 2p are found to have one of three orientations in the applied field and so interact with it differently (Fig. 1.2). The magnetic quantum number, m, is therefore a description of the orientation of the electron to the applied field.

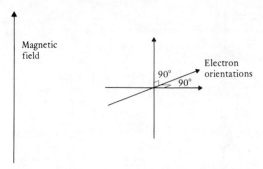

Fig. 1.2 Orientations of the 2p electrons relative to an applied magnetic field

In contrast to the 2p electrons, the 2s electrons are randomly orientated to the field and are not split into specific groups; d electrons fall into five groups and f electrons have seven different orientations. The p electrons are found to be orientated at right-angles to each other; these orientations are described by the Cartesian axes x, y and z. So, an electron with a 2p term may be designated as $2p_x$, $2p_y$ or $2p_z$ according to its orientation to an applied field. Table 1.2 gives recognised combinations of l and m.

Table 1.2 Combinations of the azimuthal and magnetic quantum numbers – the former represented by s, p, d, and the latter by the subscripts

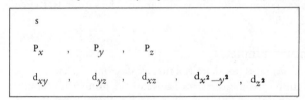

The quantum number description of an electron (e.g. $2p_x$) is called an **orbital.** These orbitals are represented diagrammatically so as to show the most probable positions of an electron of specified energy. (Since, according

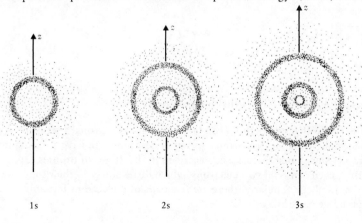

1s   2s   3s

6

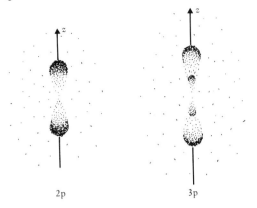

2p   3p

Fig. 1.3 Probability descriptions of some electrons

to Heisenberg's uncertainty principle, the electron's position cannot be defined exactly, all we can do is to describe its most probable positions.) Figure 1.3 illustrates some of the probability distributions – i.e. relative probabilities of the electron's positions. The more dense areas indicate positions of maximum probability of finding an electron; so the electron of energy 1s is most likely to be found in the region indicated by the densely populated area. These diagrams are cross-sections; the three-dimensional picture is obtained by rotation about the z axis shown. The s orbitals, therefore, give spherical distributions.

(d) The spin quantum number

For each of the above orbital types there is a further description. Two electrons in a single atom can have the same values of the principal, azimuthal and magnetic quantum numbers, but they must differ in the value of the spin quantum number. This term can be considered as distinguishing between the rotations of the electrons (Fig. 1.4).

Fig. 1.4 Representations of the spin quantum numbers

It is found that each orbital can hold up to two electrons; a set of p orbitals can hold six electrons and a group of d orbitals can have a maximum of ten electrons. Since there are six electrons in the 2p set of orbitals ($2p_x$, $2p_y$, $2p_z$), each can hold two electrons, which differ solely in their spin quantum numbers. Similarly, there are five pairs of d electrons but only one pair of s electrons.

Aufbau principle

When an atom contains several electrons, they are assigned to orbitals according to the following rules, which are known collectively as the 'Aufbau principle'.

(a) The electron must occupy the orbital of lowest energy, that is, the orbital representing the strongest attractive force by the nucleus. The order is 1s, 2s, 2p, 3s, 3p, 4s, 3d etc.

(b) The electrons must differ in at least one of the four quantum numbers. For example, two electrons may have the description $2p_x$, but they must be of opposite spin. This is shown diagrammatically as in Fig. 1.5. The

$2p_x$

Fig. 1.5 Representation of two electrons in the orbital $2p_x$

box represents the $2p_x$ orbital; the arrows correspond to the two spin quantum numbers. This is known as Pauli's principle.

(c) In a set of degenerate orbitals (that is, orbitals of the same energy), the electrons occupy each orbital singly before any pairing occurs, so that the atom has the maximum number of parallel spins in these orbitals. By the description 'parallel spins' we mean spins which are represented by arrows pointing in the same direction in the box notation. Degenerate orbitals occur when the orbitals differ only in their magnetic quantum numbers; $2p_x$, $2p_y$ and $2p_z$ are a set of degenerate orbitals. So, three electrons in the 2p orbitals would be arranged thus

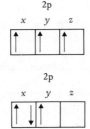

rather than as

This is known as Hund's rule.

Electronic configuration

Using the rules described above, it is possible to determine the electronic configuration of an element. Figure 1.6 shows the ordering of the orbitals involved in the elements of atomic numbers 1 to 30. Consider an element with 25 electrons. The electrons are introduced into the lowest orbital (rule (a) above), and paired in accordance with Pauli's principle. The structure is

$1s^2 \ 2s^2 \ 2p^6 \ 3s^2 \ 3p^6 \ 4s^2 \ 3d^5$

8

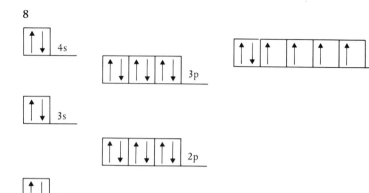

Fig. 1.6 Order of the orbitals

(The superscripts indicate the number of electrons in the preceding orbital.)
The five 3d electrons will be distributed in accordance with Hund's rule:

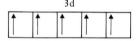

Ionisation energies

The energy required to remove an electron from an atom is known as the
ionisation energy. When an atom contains a number of electrons, these can be
removed successively. For example, the successive ionisation energies of
oxygen are listed in Table 1.3. The values are plotted in Fig. 1.7. A logarithmic

Table 1.3 The minimum molar energy required to remove the nth
successive electron from an oxygen atom

n	1	2	3	4	5	6	7	8
I.E./kJ mol^{-1}	1 310	3 400	5 300	7 500	11 000	13 300	71 300	84 100

scale is used to condense the results into a more compact space. The electronic
configuration of oxygen is

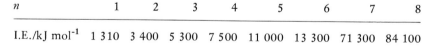

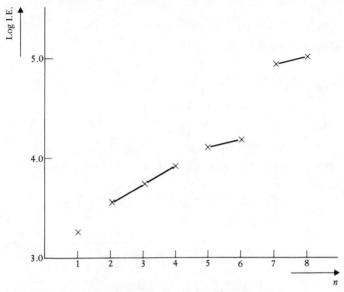

Fig. 1.7 Successive ionisation energies of oxygen

The first electron removed is the one with the least attractive force to the nucleus; it is the paired electron in the 2p set of orbitals. The three electrons in the 2p orbitals with identical spin quantum numbers are then removed successively, followed by the 2s electrons. The highest ionisation energies correspond to the 1s electrons. There is always a large energy step between the p orbitals and the s orbitals of the next higher principal quantum number, e.g. 3s and 2p, 4s and 3p, etc. A similar large jump occurs between 1s and 2s, since there is no 1p orbital.

Periodic classification

The elements can be tabulated according to their electronic structures. For example, the elements with an s^1 structure are

hydrogen	$1s^1$
lithium	$2s^1$
sodium	$3s^1$
potassium	$4s^1$
rubidium	$5s^1$
caesium	$6s^1$
francium	$7s^1$

Similarly, the elements with the structure p^1 are

boron	$2p^1$
aluminium	$3p^1$
gallium	$4p^1$

indium $5p^1$
thallium $6p^1$

The p^6 formulation includes

neon $2p^6$
argon $3p^6$
krypton $4p^6$
xenon $5p^6$
radon $6p^6$

A table (Table 1.4) can be drawn up showing these relationships. This is known as the **periodic table**. The table drawn up in this manner consists of

Table 1.4   Periodic table

| | | | | | | | | | | | 1 H | 2 He |

s^1	s^2										p^1	p^2	p^3	p^4	p^5	p^6
3 Li	4 Be										5 B	6 C	7 N	8 O	9 F	10 Ne
11 Na	12 Mg	d^1	d^2	d^3	d^4	d^5	d^6	d^7	d^8	d^9	d^{10}	13 Al	14 Si	15 P	16 S	17 Cl	18 Ar
19 K	20 Ca	21 Sc	22 Ti	23 V	24 Cr	25 Mn	26 Fe	27 Co	28 Ni	29 Cu	30 Zn	31 As	32 Ga	33 Ge	34 Se	35 Br	36 Kr
37 Rb	38 Sr	39 Y	40 Zr	41 Nb	42 Mo	43 Tc	44 Ru	45 Rh	46 Pd	47 Ag	48 Cd	49 In	50 Sn	51 Sb	52 Te	53 I	54 Xe
55 Cs	56 Ba	57 La *	72 Hf	73 Ta	74 W	75 Re	76 Os	77 Ir	78 Pt	79 Au	80 Hg	81 Tl	82 Pb	83 Bi	84 Po	85 At	86 Rn
87 Fr	88 Ra	89 Ac †	104 Rf	105 Ha	106												

	f^1	f^2	f^3	f^4	f^5	f^6	f^7	f^8	f^9	f^{10}	f^{11}	f^{12}	f^{13}	f^{14}
*	58 Ce	59 Pr	60 Nd	61 Pm	62 Sm	63 Eu	64 Gd	65 Tb	66 Dy	67 Ho	68 Er	69 Tm	70 Yb	71 Lu
†	90 Th	91 Pa	92 U	93 Np	94 Pu	95 Am	96 Cm	97 Bk	98 Cf	99 Es	100 Fm	101 Md	102 No	103 Lr

the elements in order of increasing atomic number (see below). Elements in the same column have similar chemical properties (as well as similar electronic arrangements), though there is frequently a gradual change in these properties on descending a group (see Ch. 6). Hydrogen and helium are classified separately because their small sizes and the absence of p orbitals for $n = 1$ give slightly anomalous results in terms of their chemical properties. (For reasons beyond the theory dealt with in this volume, a few of the configurations deviate from those implied in Table 1.4, but they do not affect the general arguments used in this book.)

Nuclear structure

The nucleus is not involved in normal chemical reactions and so its structure is not considered in detail. Nuclear chemistry is involved in radioactivity (see the author's *Fundamentals of Chemistry*). The main nuclear particles are the proton and the neutron. Each element has a specific number of protons. This is the **atomic number**. For example, magnesium has the atomic number 12, so it must contain twelve protons in its nucleus. As shown in Table 1.5, the proton

Table 1.5 Properties of the fundamental atomic particles

	Mass/kg	Relative mass/ atomic mass units*	Charge/C	Relative charge
Proton	$1.672\ 52 \times 10^{-27}$	1	$+1.602\ 10 \times 10^{-19}$	$+1$
Neutron	$1.674\ 82 \times 10^{-27}$	1	0	0
Electron	$9.109\ 08 \times 10^{-31}$	0	$-1.602\ 10 \times 10^{-19}$	-1

*The 'atomic mass unit', amu, is sometimes described as a dalton.

and electron have charges of the same magnitude but of opposite sign. Therefore, *in a neutral atom*, there must be equal numbers of protons and electrons.

The number of neutrons is less critical to the chemical nature of the atom. For example, magnesium consists of a mixture of atoms with varying numbers of neutrons (Table 1.6).

Table 1.6 Nuclear structure of the isotopes of magnesium

Total number of nuclear particles (mass number)	Percentage abundance/ %	Number of protons (atomic number)	Number of neutrons	Isotopic mass/amu
24	78.60	12	12	23.9925
25	10.11	12	13	24.9938
26	11.29	12	14	25.9898

Atoms which possess the same number of protons but differ in their number of neutrons are known as *isotopes*.

Since the nucleus contains most of the mass of the atom (see Table 1.5) and the protons and neutrons are of approximately equal relative mass, the total number of nuclear particles ('nucleons') is known as the **mass number.** As is apparent from Table 1.6, the mass number deviates slightly from the actual relative mass (the **isotopic mass**) of the atom. For most purposes this deviation is not significant and can be overlooked, but when modern instruments of high precision, such as the mass spectrometer, are used, the accurate values are more suitable.

A careful examination of the values listed in Table 1.6 will reveal that the isotopic mass is slightly less than that expected from the masses of the constituent particles. This loss of mass is due to the energy released from the nucleus in binding the nucleons together (the 'binding energy'). It is found that the percentage composition of a mixture of isotopes of a single element (e.g. magnesium in Table 1.6) is often constant regardless of the primary source of that material. This means that it is possible to quote the **relative atomic mass** of the element, a value which takes into account the percentage

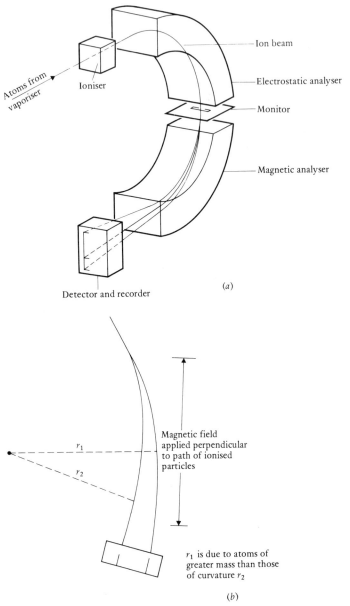

Atoms from vaporiser

Ioniser

Ion beam

Electrostatic analyser

Monitor

Magnetic analyser

Detector and recorder

(a)

r_1

r_2

Magnetic field applied perpendicular to path of ionised particles

r_1 is due to atoms of greater mass than those of curvature r_2

(b)

Fig. 1.8 Diagrammatic representation of (a) a high resolution mass spectrometer, showing (b) the effect of magnetic field

isotopic composition. It is the weighted average of the isotopic masses. This can be illustrated for magnesium. The relative atomic mass (m_r) is given by

$$m_r = \left(\frac{78.60}{100} \times 23.992\,5\right) + \left(\frac{10.11}{100} \times 24.993\,8\right) + \left(\frac{11.29}{100} \times 25.989\,8\right)$$

$$= 24.321\,2$$

The values of the isotopic masses and percentage compositions are determined using a high resolution mass spectrometer (Fig. 1.8a). The element is vaporised in an oven and is then passed into the ioniser, in which the atoms are bombarded with high energy electrons. The electrons collide with the atoms and positive ions are formed (*Fundamentals of Chemistry*, p. 39). The ionised particles are drawn as a beam into the electrostatic field in which they are accelerated to give them a greater kinetic energy. The kinetic energy, $\frac{1}{2}mv^2$, is equal to $e \cdot V$, where e is the electronic charge and V is the electric potential. The charged particles are then passed into a magnetic field which deflects them. The radius of curvature, r, for the deflection is dependent on the mass and velocity of the particles in accordance with the equation

$$\frac{m \cdot v^2}{r} = H \cdot e \cdot v$$

where H is the intensity of the magnetic field. Combining these equations, we find that the radius of curvature (see Fig. 1.8 b) is given by

$$r = \sqrt{\frac{2\,mV}{e \cdot H^2}}$$

If several isotopes are present, then these will be focused at different positions on the detector. Figure 1.9 gives the spectrum for potassium. The

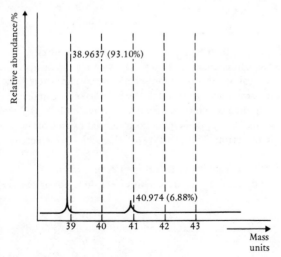

Fig. 1.9 Mass spectrum of potassium

relative intensities of the lines are proportional to the relative abundances of the isotopes in the element.

Metallic bonding

The structures of covalent and ionic compounds are described in *Fundamentals of Chemistry* and the distinctive properties of these species are discussed. Metallic bonding cannot be explained in either of these ways since the properties of metals do not correspond with either of these types of bonding (see Table 1.7).

Table 1.7 Characteristics of the main types of bonding

	Ionic compounds	Covalent compounds	Metals
Physical state	Solids of high melting point; brittle, hard	Liquids or solids of low melting point; soft	Solids, usually of high melting point; generally hard, strong and ductile
Conductivity	Non-conductors in the solid state; conductors in the molten state	Non-conductors in all states	Conductors in all states
Solubility	Soluble in polar solvents	Soluble in non-polar solvents	Generally only dissolve in polar solvents by reaction; soluble in liquid metals

The bonding in metals can be described simply as follows. The outer orbitals of the metals overlap as in covalent bonding. The valency electrons (the electrons occupying the orbitals of lowest ionisation energy) are then free to move randomly from atom to atom through the interlinked orbitals. The resulting structure corresponds to a lattice of metal ions which is permeated by a sea of electrons. An applied electrical potential causes these electrons to move in a single direction. This is a flow of electricity, an electric current.

In ionic solids, a shear force causes crystal cleavage because of the repulsion between like charged ions (Fig. 1.10 a). This effect does not arise

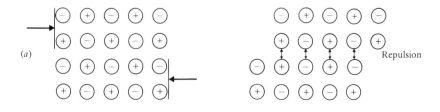

(a)

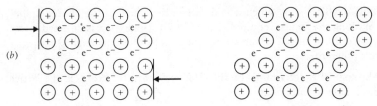

Fig. 1.10 Effects of shear forces on (a) ionic solids and (b) metals

in metals because of the binding effect of the mobile electrons (Fig. 1.10 b).

Crystal structures

X-ray diffraction

When monochromatic X-rays (that is, X-rays of a single wavelength) are focused on to a crystal, the rays are diffracted by the atoms in the crystal. The pattern created by the diffracted rays is determined by the arrangement of the particles in the crystal lattice. The crystal is a three-dimensional structure. In order to examine all the planes of the crystal, the diffraction pattern is obtained either for a rotating crystal or for a powder of the solid. In the latter case, due to the random orientations of the particles, all aspects of the lattice are considered. The basic arrangement of an X-ray diffracto-meter is shown in Fig. 1.11.

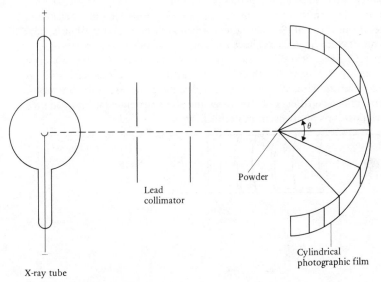

Fig. 1.11 Diagrammatic representation of a diffractometer

The X-rays are generated by firing a beam of electrons on to a metal target; this causes the excitation of the metal atoms which then lose the excess energy as X-radiation. The lead collimators are used to prevent the spreading of the X-rays, which would darken the whole of the photographic film, masking the results required. The beam of X-radiation is directed on to the crystal, the atoms of which scatter the rays through the angle θ, or multiples thereof.

The angle θ is determined by the arrangement of the atoms in the lattice. Bragg was able to show that

$$n \cdot \lambda = 2 \cdot d \cdot \sin \theta$$

where λ is the wavelength of the X-radiation;

d is the distance between identical layers of atoms;

θ is the angle shown in Fig. 1.11, and is the angle of incidence of the radiation to this set of planes of atoms;

n is an integer from 1 to infinity, though usually with the lower values.

For example, for a potassium chloride crystal, X-rays of wavelength 153.7 pm are first reflected (i.e. $n = 1$) at an angle of $14°\,9'$. Substitution in the Bragg equation gives

$$1 \times 153.7 \times 10^{-12} = 2 \times d \times \sin 14°9'$$

$$\therefore d = \frac{153.7 \times 10^{-12}}{2 \times 0.25}$$

$$= 3.14 \times 10^{-10}\,m$$

So, the distance between the planes is 314 pm.

It can be seen that the formula does not take into account the chemical nature of the material. So, substances which are isomorphous (that is, have the same crystal dimensions) have the same X-ray diffraction patterns. Figure 1.12 gives some diffraction patterns for powders. A stationary crystal gives a series of dots, rather than lines, the positions of these being similarly related to internuclear distances. This technique provides a valuable tool for the determination of crystal structures and so to the elucidation of the relative positions of atoms in molecules.

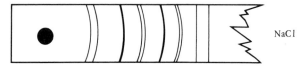

NaCl

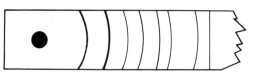

KCl

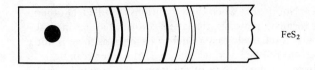

FeS$_2$

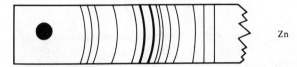

Zn

Fig. 1.12   Portions of some powder diffraction patterns

Metallic crystal structures

Crystals have fascinated man for many generations. They have been used as objects of beauty by craftsmen and artists alike. If possible, obtain some crystals and examine their characteristics. One of the notable features is the constancy of shape in the crystals of a single substance. This can be explained as resulting from a regular arrangement of the particles (normally atoms or ions) in the solid lattice. Draw a rectangle 86 × 46 mm. How many ½p coins can be fitted into this rectangle? If you use the arrangement as in Fig. 1.13,

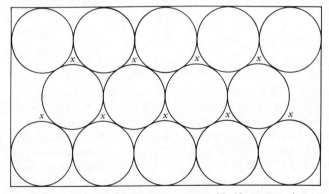

Fig. 1.13   Close-packing arrangement of half-penny coins

you should be able to pack fourteen into it. You will find that any other arrangement holds fewer coins. So this assembly is known as a close-packing arrangement. A similar system is used for packing spherical objects such as apples. The same principle is used in three dimensions. Arrange a close-packed layer of 'plastic' spheres in a tray. Now introduce another layer of spheres on top of this one. Where are they most stable? In the gaps between the spheres – for example, in the positions indicated by x in the two-dimensional Fig. 1.13. Notice that when these positions are filled, another close-packed layer is formed. In metals, the layers consist of atoms in close-packed arrangements. It is relatively easy to deform a metal as there is not a large barrier to movement of one layer over another. Movement can be hindered by the inclusion of small atoms in the spaces unoccupied by the metal atoms. Typical positions are those gaps left unmarked in Fig. 1.13. For example, iron is less easily worked after the inclusion of carbon to form steel. The carbon atoms hinder relative movement of layers. Similar principles apply in other alloy mixtures. (This is not a complete explanation of alloying properties, but it illustrates one effect.)

Packing arrangements

There are two ways in which a symmetrical close-packing arrangement can be achieved three-dimensionally. Figure 1.14 illustrates the first two layers in a close-packing structure. The third layer of spheres can be placed in one

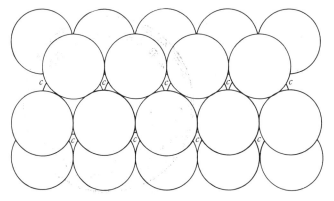

Fig. 1.14   Two layers in a close-packing arrangement

of two possible positions: either directly above the first layer, or over the positions not yet occupied by any spheres (positions c in Fig. 1.14). If the first layer is designated as A, the second layer as B, then the symmetrical arrangements arising from these two systems can be denoted as ABABAB . . . and ABCABC . . . The first configuration is known as the **hexagonal close-packing arrangement** and the second is described as a **cubic close-packing arrangement**. These names specify the symmetry arising from these structures (see Fig. 1.15). The hexagonal close-packing arrangement has a 'six-fold axis of symmetry' (Fig. 1.15 a). In the latter configuration, the highest degree of symmetry is that of a cube with a threefold axis of symmetry (Fig. 1.15 b). A sixfold axis of symmetry means that when the structure is rotated about this axis, the same arrangement of the atoms occurs six times during a 360° rotation.

The common packing arrangements of some metals are given in Table 1.8. It will be noticed that some metals have more than one possible configuration. This characteristic of having alternative structures is known

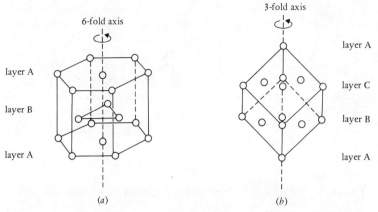

Fig. 1.15   Symmetry descriptions of the close-packing arrangements

as **allotropy**. So, the two allotropes of calcium have cubic close-packed and
hexagonal close-packed lattices.

Table 1.8   Some metals with close-packing structures

C.C.P.		H.C.P.	
Ca	Ni	Mg	Co
Sr	Cu	Ca	Ni
Fe	Ag	Ti	Zn
Co	Au	Mo	Cd

There is a further structural type that is found in metals: **body-centred
cubic arrangement**. There is an atom at each corner of the cube and one in
the centre of the unit. This is not a close-packed arrangement, since it uses
only 68 per cent of the available space in contrast to 74 per cent in the close-
packing systems. Typical examples of body-centred cubic structures are to be
found in sodium, potassium, titanium, vanadium, chromium and iron.

Ionic structures

In ionic compounds we consider the larger ions in the close-packed arrangement. The smaller ions occur in the smaller sites as described for carbon in
iron. For example, consider sodium chloride. The ionic radii are Na^+ 95 pm,
Cl^- 181 pm. The Cl^- ions are close-packed. In the centre of the cubic structure
(Fig. 1.16 a), between layers B and C is a space which can be occupied by a
sodium ion. This sodium ion acts as a centre of attraction for the neighbouring negative ions, so overcoming their strong repulsion for each other. The ion
is a little too large for the site and so there is some displacement of the
chloride ions from their nominal positions. The positions of the other sodium
ions are as shown in Fig. 1.16 b. It will be seen that the sodium ions are in the
spaces between the chloride ions (the mid-points of the edges of the cube).

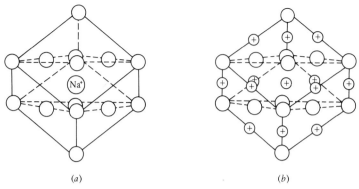

Fig. 1.16 The position of sodium ions in a sodium chloride lattice

Consider the position of a sodium ion relative to the chloride ions with which it is in contact. Figure 1.17 *a* shows an extract from Fig. 1.16 *a*; Fig. 1.17 *b* illustrates that the sodium is in an octahedral position (or 'site') with respect to the chlorides. All the sodium ions are in these octahedral

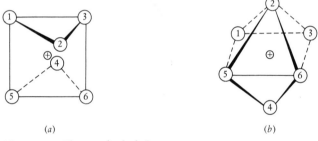

Fig. 1.17 The octahedral site

sites, and all the octahedral sites are filled. There is one octahedral site for each atom in the lattice. So, the sodium chloride structure can be described as a cubic close-packing of the chloride ions, with all the octahedral sites filled, and there are equal numbers of Na^+ and Cl^- ions.

There is an alternative type of site in the structures. Figure 1.18 shows

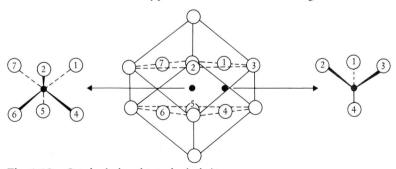

Fig. 1.18 Octahedral and tetrahedral sites

the presence of the octahedral and tetrahedral sites in a close-packed lattice. There are twice as many tetrahedral sites as octahedral sites. These sites can be occupied instead of the octahedral positions. For example, if calcium ions are in a cubic close-packing arrangement, fluoride ions can occupy the tetrahedral sites, giving the structure $Ca^{2+}F_2^-$ (Fig. 1.19).

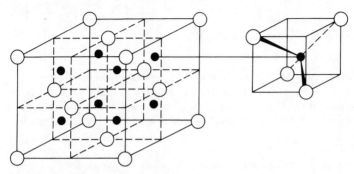

Fig. 1.19   The fluorite structure

If the sulphide ions are in a cubic close-packing arrangement and zinc ions occupy *half* the tetrahedral sites, then $Zn^{2+}S^{2-}$, zinc blende, results (Fig. 1.20).

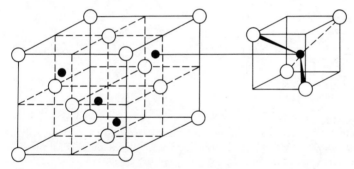

Fig. 1.20   The zinc blende structure

Many other ionic lattice types are known, but these three describe the basic principles which can be extended to other situations.

It was mentioned above that metals can be deformed by the shift of one layer over another − there is little hindrance to such movement. In the ionic lattice we have two new factors − the spaces between layers are filled, so hindering movement (compare this with the comment on alloys), and when a layer is displaced we see that this causes ions of like charge to be closer to each other (causing a strong repulsion). So, ionic crystals are hard (due to the attractive forces) and they cleave easily when a shear force is applied to them (Fig. 1.21). A large shear force is required to cause movement because of steric hindrance, but once displacement is achieved the repulsive forces take effect.

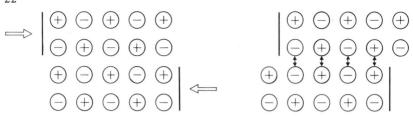

Fig. 1.21 Shear forces in an ionic crystal

Effect of covalency

Even in ionic compounds there is often a significant degree of covalent character (see *Fundamentals of Chemistry*, p. 37). It is found that, whereas ionic compounds of the form A^+B^- tend to assume a sodium chloride structure, often those with substantial covalent nature have a zinc blende structure.

Covalent structures

Crystals based on non-polar covalent molecules (for example, iodine) are generally weak, very soft and of low melting point. The iodine crystal consists of diatomic iodine molecules packed in a manner similar to the close-packing arrangement (Fig. 1.22). Movement is easy in each direction since there are only weak forces between the molecules (see *Fundamentals of Chemistry*, p. 38).

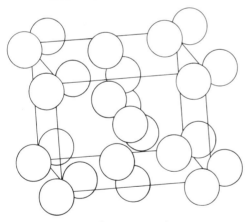

Fig. 1.22 Crystal structure of iodine

Polar covalent molecules crystallise with similar arrangements to those of the crystals of non-polar molecules but are harder and less volatile due to the larger intermolecular forces. These arise from attractions between centres of polarity in adjacent molecules (Fig. 1.23). This is illustrated in ice (Fig. 1.24); the oxygen atoms are in an approximately close-packing arrangement and are separated by hydrogen atoms. The water molecules are held in place by strong polar forces between oxygen atoms of one molecule and hydrogen atoms of another.

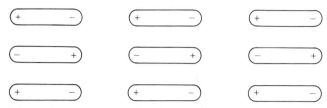

Fig. 1.23 Polar intermolecular forces

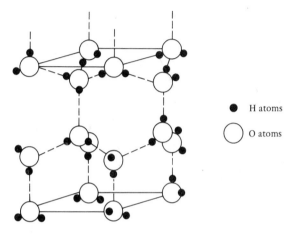

H atoms

O atoms

Fig. 1.24 Crystal structure of ice

In some covalent structures a continuous two- or three-dimensional network of bonds is set up. Graphite consists of layers of linked carbon atoms; diamond involves the same atoms in a three-dimensional array (Fig. 1.25).

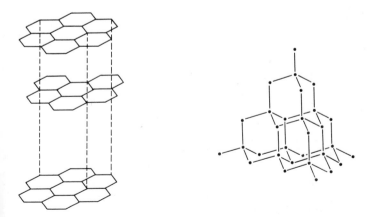

Fig. 1.25 Crystal structure of (*a*) graphite and (*b*) diamond

Table 1.9 gives a comparison of some typical crystal types and their main properties.

Table 1.9   General relationships between structure and properties

	Monatomic	Non-polar molecular	Polar molecular	Giant molecular	Ionic lattice	Metallic
Particle type	Single atoms	Non-polar covalent molecules	Polar covalent molecules	Polymeric units	Ions	Atoms
Bonding	Weak inter-atomic	Covalent and weak inter-molecular	Covalent and electro-static inter-molecular	Covalent between units; weak between sheets	Electro-static	Metallic with delocalised electrons
Electrical conductance (a) *solid* (b) *liquid*	Zero Zero	Zero Zero	Very low Very low	Zero Zero	Zero High	Very high Very high
Solubility in water	Insoluble	Insoluble	Variable	Insoluble	Variable	Insoluble
Solubility in covalent solvents	Soluble	Soluble	Soluble	Insoluble	Insoluble	Insoluble
Physical state	Gaseous	Volatile liquids or soft, weak crystals	Soft crystals, little strength and low m.p.	Hard crystals, variable strength and high m.p.	Hard, brittle crystals with high m.p.	Solids; malleable, variable strength; usually high m.p.

Lattice energy

Hess' law (*Fundamentals of Chemistry*, p. 166) states that the total energy change in a process is independent of the path taken in that process; it is determined solely by the energy of the initial reactants and the final products. This can be applied to the formation of crystalline compounds; for example, sodium chloride:

$$Na(s) + \tfrac{1}{2}Cl_2(g) \rightarrow Na^+Cl^-(s)$$

The enthalpy change for this reaction is known as the enthalpy of formation, ΔH_f. This overall change can be broken down into several steps:

atomisation, $\quad Na(s) + \tfrac{1}{2}Cl_2(g) \rightarrow Na(g) + \tfrac{1}{2}Cl_2(g), \; \Delta H_a$

dissociation, $\quad Na(g) + \tfrac{1}{2}Cl_2(g) \rightarrow Na(g) + Cl(g), \quad \Delta H_d$

ionisation, $\quad Na(g) + Cl(g) \rightarrow Na^+(g) + Cl(g), \quad \Delta H_i$

anion formation, $Na^+(g) + Cl(g) \rightarrow Na^+(g) + Cl^-(g)$, ΔH_e

lattice formation, $Na^+(g) + Cl^-(g) \rightarrow Na^+Cl^-(s)$, ΔH_l

$$\Delta H_f = \Delta H_a + \Delta H_d + \Delta H_i + \Delta H_e + \Delta H_l$$

These changes can be represented as a cycle, known as the Born–Haber cycle (Fig. 1.26).

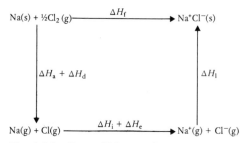

Fig. 1.26   Born–Haber cycle

This cycle can be used to estimate unknown values; for example, ΔH_f values, if all the other terms are known. The factors ΔH_a, ΔH_d, ΔH_i and ΔH_e are usually readily available from data tables, and ΔH_l can be estimated by the use of formulae derived on the basis of electrostatic attractions.

Calcium chloride ($Ca^{2+}Cl^-_2$) has an enthalpy of formation of -795 kJ mol^{-1}, but what is the value for Ca^+Cl^-? The lattice energy for this is estimated to be -722 kJ mol^{-1}, and the other figures are

$\Delta H_a = +172$ kJ mol^{-1}
$\Delta H_i = +589$ kJ mol^{-1}
$\Delta H_d = +124$ kJ mol^{-1} (i.e. per mole of Cl atoms)
$\Delta H_e = -348$ kJ mol^{-1}

Substitution in the above equation gives ΔH_f for Ca^+Cl^- as -185 kJ mol^{-1}.

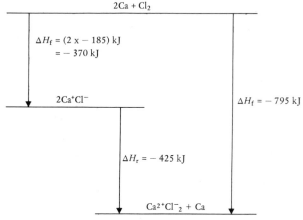

Fig. 1.27   Energy level diagram for calcium chloride

While this would be expected to be a stable structure, it is seen that $Ca^{2+}Cl^-_2$ is more stable. In fact, construction of an energy diagram (Fig. 1.27) shows that the change $2Ca^+Cl^-(s) \rightarrow Ca^{2+}Cl^-_2(s) + Ca(s)$ has an enthalpy change of -425 kJ mol^{-1}.

Summary

At the conclusion of this chapter, you should be able to:

1. state why it is difficult to define both the position and the momentum of an electron simultaneously;
2. state Heisenberg's uncertainty principle;
3. state the premises on which Schrödinger based his theory;
4. name the four quantum numbers;
5. state that the number of values of the azimuthal quantum number is equal to the value of the principal quantum number;
6. explain what is meant by an orbital;
7. state and apply the Aufbau principle;
8. write the electronic structure of all the elements with atomic numbers 1 to 30;
9. define ionisation energy;
10. relate the periodic table classification to the electronic structure;
11. state that the atomic number is the number of protons in the atomic nucleus;
12. define isotopes and mass number;
13. calculate the relative atomic mass of an element, given its isotopic composition;
14. describe the essential components of a mass spectrometer;
15. explain simply the nature of metallic bonding;
16. relate the physical properties of a metal to its metallic bonding;
17. describe the formation of an X-ray diffraction pattern for a crystalline solid;
18. describe the component parts of an X-ray diffractomer;
19. calculate the distance between atomic planes, given the angle of diffraction and wavelength;
20. describe a close-packing arrangement in a metallic crystal;
21. distinguish between hexagonal and cubic close-packing arrangements;
22. define allotropy;
23. state that a body-centred cubic arrangement is not a close-packing system;
24. relate ionic structures to the basic packing arrangements;
25. recognise the sodium chloride, calcium fluoride and zinc sulphide structures;
26. relate the physical properties of ionic compounds to their crystal structures;
27. describe some typical covalent crystal structures;
28. describe the components of the Born—Haber cycle;

29. perform calculations based on the Born–Haber cycle;
30. show why, for example, $Ca^{2+}Cl^-_2$ (s) is more stable than Ca^+Cl^-(s).

Experiments

1.1 Determination of the ionisation energy of a noble gas

Set up a circuit as shown in Fig. 1.28.

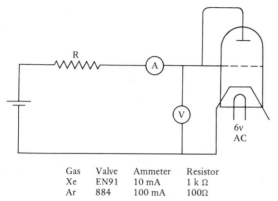

Gas	Valve	Ammeter	Resistor
Xe	EN91	10 mA	1 k Ω
Ar	884	100 mA	100Ω

Fig. 1.28   Circuit for the ionisation energy of a noble gas

Increase the voltage in 2 V steps up to 10 V and then by smaller increments. At each step note the ammeter reading and tabulate the voltage and current values. The maximum ammeter readings used should be 8 mA for xenon and 80 mA for argon. Plot your results. When ionisation occurs, there is a more rapid increase in the current flow. Determine the ionisation potential for the gas. Convert the ionisation potential to an ionisation energy (kJ mol^{-1}). (For a discussion of the operation of the valve, see *Nuffield Advanced Science, Chemistry, Student's Book I* (Longman, 1970), pp. 78–82.)

1.2 Cleavage of covalent and ionic crystals

Take a large piece of a crystalline solid (e.g. calcite). Examine the overall shape, including imperfections. Tap the crystal gently with a blunt instrument (e.g. a hammer). Examine the shape of the pieces and the manner of fragmentation of the original lump. Compare the results for an ionic crystal and a covalent crystal. Relate the observations to the structures described in the text.

1.3 Construction of crystal models

Using polystyrene spheres of appropriate scale sizes: (*a*) construct several close-packed layers; and (*b*) assemble these layers to produce the two types of close packing.

Locate the octahedral and tetrahedral sites. Construct typical lattice structures for ionic solids. (A suitable guide for this work is *Making Crystal Models* by R. E. Darby (Pergamon, 1969).)

28

References

A New Guide to Modern Valency Theory; G. I. Brown (Longman, 2nd edn., 1967).
Crystals and Crystal Structure; M. Hudson (Longman, 1971).
Allotropy; W. E. Addison (Oldbourne, 1964).
X-ray diffraction, *Pye-Unicam Analytical Bulletin Supplement*, 1978.
Chemical bonds (4 parts); N. Booth, *Educ. Chem.*, 1964, 1, 16–24, 66–74, 151–6, and 210–17.

Films

Chemical bonding (CHEM Study).
The hydrogen atom (CHEM Study).
Considering crystals (Unilever).
Crystal structure (ICI).
X-ray crystallography (General Electric Co.).

Questions

1. Summarise the principles underlying Schrödinger's description of an electron and the consequences of his interpretation.
2. Describe the electronic arrangements of the following elements (atomic numbers given): (a) $_5$B; (b) $_{17}$Cl; (c) $_{12}$Mg; (d)$_{24}$Cr; (e) $_{18}$Ar.
3. The table below gives the successive ionisation energies of fluorine. Plot a graph of the log(I.E.) against the number of the electron removed and relate the form of the graph to the electronic structure.

I.E./kJ mol^{-1}	1 680	3 400	6 000	8 400	11 000	15 200	17 900	92 000	106 000
Electron no.	1	2	3	4	5	6	7	8	9

4. Iron has the following isotopic composition: 5.81 per cent ^{54}Fe; 91.64 per cent ^{56}Fe; 2.21 per cent ^{57}Fe; 0.34 per cent ^{58}Fe. The isotopic masses are respectively 53.957 5, 55.953 1, 56.954 6 and 57.950 6. Calculate the relative atomic mass of iron.
5. Draw the crystal structure of sodium chloride and discuss: (a) the close-packing arrangement of the Cl$^-$ ions; and (b) the position of the Na$^+$ ions in the lattice.
   Describe the physical properties of sodium chloride and explain them in terms of the ionic lattice described.
6. From the data supplied, calculate the enthalpy of formation for Mg$^+$Cl$^-$(s). Account for the fact that this compound is not formed.

Process	ΔH/kJ mol^{-1}
$Mg(s) \rightarrow Mg(g)$	+141
$Mg(g) \rightarrow Mg^+(g) + e^-$	+732
$\frac{1}{2}Cl_2(g) \rightarrow Cl(g)$	+124
$Cl(g) + e^- \rightarrow Cl^-(g)$	−348
$Mg^+(g) + Cl^-(g) \rightarrow Mg^+Cl^-(s)$	−808
$Mg(s) + Cl_2(g) \rightarrow Mg^{2+}Cl^-_2(s)$	−642

Chapter 2

Equilibrium and energetics

Background reading

The background reading to this chapter is *Fundamentals of Chemistry*, Chapters 13 and 14.

Acids and bases

Acids are substances that, in solution, generate solvated hydrogen ions (see *Fundamentals of Chemistry*, p. 29). For example, if a substance HA dissolves in a solvent X in accordance with the equation

$$HA + X \rightleftharpoons XH^+ + A^-$$

then HA is an acid since it produces the solvated hydrogen ions, XH^+. If the solvent is liquid ammonia, then the solvated hydrogen ion is NH_4^+, the ammonium ion. Water solvates the hydrogen ion to give the hydronium ion, H_3O^+. The latter ion is sometimes represented as $H^+(aq)$. Hydrogen chloride is a covalent gas which dissolves in water to produce hydronium ions. It is, therefore, an acid.

$$HCl + H_2O \rightarrow H_3O^+ + Cl^-(aq)$$

Similarly, ethanoic acid (CH_3COOH) is a covalent liquid at 293 K. It forms an acidic solution in water.

$$CH_3COOH + H_2O \rightleftharpoons CH_3COO^-(aq) + H_3O^+$$

J. N. Bronsted and J. M. Lowry extended this classification further to say that acids are substances that donate hydrogen ions in a reaction. This definition includes the above examples in which the hydrogen chloride and ethanoic acid molecules donate hydrogen ions to the solvent. It also applies to a reaction such as

$$HCl + OH^- \rightarrow H_2O + Cl^-$$

in which the hydrogen chloride donates H^+ ions to the hydroxide ions. Identify the acid in the following reactions:

(a) $HNO_3 + H_2O \rightarrow H_3O^+ + NO_3^-$
(b) $C_2H_5OH + HCl \rightleftharpoons C_2H_5OH_2^+ + Cl^-$
(c) $H_2SO_4 + Cl^- \rightarrow HCl + HSO_4^-$
(d) $NH_3 + HCl \rightarrow NH_4^+ + Cl^-$
(e) $CH_3COO^- + H_2O \rightleftharpoons CH_3COOH + OH^-$

In each case we must identify the substance that donates a hydrogen ion. This substance can be easily recognised by a loss of H^+ from its formula. The HNO_3 molecule in (a) loses H^+ to become NO_3^- and so is the acid. In (b), it is the hydrogen chloride (HCl) that loses H^+. Sulphuric acid and hydrogen chloride in (c) and (d) respectively lose H^+. In (e), water loses H^+ and so behaves as an acid.

Bases are similarly defined as substances which accept protons (hydrogen ions). Hydroxide is a base in the reaction with hydrogen chloride

$$HCl + OH^- \rightarrow H_2O + Cl^-$$

since it accepts an H^+ ion to produce water. Application of this definition to equations (a) to (e) above identifies the following species as bases: water in (a), ethanol (b), chloride ions (c), ammonia (d), and ethanoate (e).

It will be noticed that water behaves as a base in (a) and as an acid in (e). A compound which can be both acidic and basic is termed amphiprotic. Such behaviour in water is not surprising when it is recalled that this substance dissociates into hydronium ions (acidic) and hydroxide ions (basic):

$$2H_2O \rightleftharpoons H_3O^+(aq) + OH^-(aq)$$

Consider again example (e). The reaction between the ethanoate ions and water molecules produces ethanoic acid molecules and hydroxide ions. These are acidic and basic respectively. For example, they react together to give ethanoate ions and water:

$$CH_3COOH + OH^- \rightleftharpoons CH_3COO^- + H_2O$$
$$\text{acid} \qquad \text{base} \qquad \text{base} \qquad \text{acid}$$

In (e), CH_3COOH is known as the **conjugate acid** of CH_3COO^-, and OH^- is the **conjugate base** of water.

This can be applied to the reaction between hydrogen chloride and water:

HCl	+	H_2O	$\rightarrow$	H_3O^+	+	$Cl^-(aq)$
Acid		Base		Conjugate acid of H_2O		Conjugate base of HCl

Strengths of acids and bases

The above examples include the use of both the normal directional sign ($\rightarrow$) and the equilibrium sign ($\rightleftharpoons$) in the equations. The former implies complete reaction, and the latter indicates an incomplete reaction. An acid which reacts almost completely with water even at high concentration to produce the hydrated hydrogen ion is known as a **strong acid**. **Weak acids** react incompletely with the solvent and so set up an equilibrium. Similarly, **weak bases** react with the solvent to produce a partial dissociation into ions:

$$NH_3 + H_2O \rightleftharpoons NH_4^+ + OH^-$$

Strong bases undergo extensive ionisation.

$$NaOH + aq \rightarrow Na^+(aq) + OH^-(aq)$$

The equilibrium constant, K, for a weak acid HA

$$HA + H_2O \rightleftharpoons H_3O^+ + A^-$$

is given by

$$K = \frac{[H_3O^+]_e\,[A^-]_e}{[HA]_e\,[H_2O]_e}$$

where the symbolism $[X]_e$ denotes the equilibrium molar concentration of X. For dilute solutions of weak acids $[H_2O]_e$ is virtually constant compared to such terms as $[H_3O^+]_e$, and so the expression can be modified to

$$K_a = K \cdot [H_2O]_e = \frac{[H_3O^+]_e\,[A^-]_e}{[HA]_e}$$

and K_a is the **dissociation** (or **ionisation**) **constant** of HA.

If the acid is M molar and undergoes a degree of dissociation α (i.e. a fraction α of the molecules dissociate), then a total of $M\alpha$ mol dm^{-3} will dissociate. So, at equilibrium, $(M - M\alpha)$ mol dm^{-3} remain. Since, for every 1 mole of dissociated HA, 1 mole of H_3O^+ and 1 mole of A^- are produced, $M\alpha$ moles of HA will generate $M\alpha$ moles of both H_3O^+ and A^-. So, we can tabulate our concentrations as follows:

	HA	H_3O^+	A^-
Molar concentrations before dissociation	M	O	O
Molar concentrations at equilibrium	$(M - M\alpha)$	$M\alpha$	$M\alpha$

So, $K_a = \dfrac{(M\alpha) \cdot (M\alpha)}{(M - M\alpha)} = \dfrac{M\alpha^2}{(1-\alpha)}$

If the acid is weak, then

$$(1 - \alpha) \approx 1$$

and so,

$$K_a = M \cdot \alpha^2$$

This formula is a mathematical statement of Ostwald's law which states that for dilute solutions of weakly ionised substances the degree of dissociation is inversely proportional to the square root of the molar concentration.

For example, benzoic acid (C_6H_5COOH) has a dissociation constant 6.00×10^{-5}. What is the degree of dissociation in a 0.05 M solution? Substitution in the above expression gives the following results:

$$K_a = 6.00 \times 10^{-5}$$
$$M = 0.05$$

So, $6.00 \times 10^{-5} = 0.05 \times \alpha^2$

$$\therefore \alpha^2 = 1.2 \times 10^{-3}$$
$$\therefore \alpha = 3.46 \times 10^{-2}$$

The degree of dissociation is, therefore, 0.034 (or, 3.46%).

A similar description can be given for the dissociation of bases in solution. Ammonia is a weak base,

$$NH_3(aq) + H_2O \rightleftharpoons NH_4^+(aq) + OH^-(aq)$$

$$K = \frac{[NH_4^+]_e [OH^-]_e}{[NH_3]_e [H_2O]_e}$$

But, since this base is weak, $[H_2O]_e$ is effectively constant in dilute solutions. So,

$$K_b = K \cdot [H_2O]_e = \frac{[NH_4^+]_e [OH^-]_e}{[NH_3]_e}$$

For $NH_3(aq)$, K_b, the dissociation constant of the base, is 1.74×10^{-5}. For the ammonia solution equilibrium given above, an M molar solution of ammonia, with a degree of dissociation α, will give the following concentration terms:

	NH_3	NH_4^+	OH^-
Molarity before dissociation	M	O	O
Molarity at equilibrium	$(M - M\alpha)$	$M\alpha$	$M\alpha$

$$\therefore K_b = \frac{M\alpha^2}{(1 - \alpha)}$$

As before, $(1 - \alpha) \approx 1$ and so $K_b = M \cdot \alpha^2$

A 0.05 M solution of ammonia, therefore, has a value of α given by

$$\alpha^2 = K_b/M$$

So, $\alpha = \sqrt{\dfrac{1.74 \times 10^{-5}}{0.05}} = 1.87 \times 10^{-2}$

Ionic product of water

Water is a covalent molecule but it is slightly ionised, producing

$$2H_2O \rightleftharpoons H_3O^+ + OH^-$$

hydronium ions and hydroxide ions. The equilibrium expression for this would be

$$K = \frac{[H_3O^+]_e \, [OH^-]_e}{[H_2O]_e^2}$$

However, the concentration of hydronium ions is very small (10^{-7} mol dm^{-3} at 298 K). In contrast, the concentration of undissociated water molecules is large (55.56 mol dm^{-3}). So, even for large changes in the ionic concentrations, the relative changes in $[H_2O]$ are very small. Therefore,

$$K_w = [H_2O]_e^2 \cdot K = [H^+]_e \, [OH^-]_e$$

where K_w is the ionic product for water and is a constant if temperature is constant. Since $[H_3O^+]_e = 10^{-7} \, M$

and $[H_3O^+]_e = [OH^-]_e$, equal quantities being produced on the dissociation of water molecules,

$$K_w = 10^{-7} \cdot 10^{-7}$$
$$= 10^{-14}$$

pH scale

The strength of an acid is determined by its degree of dissociation to produce solvated hydrogen ions. The strong acids have high values for $[H_3O^+]$; weak acids have relatively low values. A convenient scale which gives us a measure of hydronium ion concentration (and so of acid strength) is the pH scale.

$$pH = -\log_{10} [H_3O^+]$$

For 0.05 M benzoic acid, it is shown above that $\alpha = 3.46 \times 10^{-2}$, so

$$[H_3O^+] = M\alpha = 0.05 \times 3.46 \times 10^{-2}$$
$$= 1.73 \times 10^{-3}$$

$$pH = -\log_{10} [H_3O^+]$$
$$= -\log_{10} (1.73 \times 10^{-3})$$
$$= 2.76$$

Pure water has $[H_3O^+] = 10^{-7} \, M$ at 298 K

and so its $pH = -\log_{10} (10^{-7})$
$$= 7$$

A pH of 7 indicates a neutral solution.

Hydrochloric acid is a strong acid and undergoes complete dissociation in aqueous solution. So, a 0.05 M solution of this acid will produce 0.05 mol dm^{-3} of H_3O^+:

HCl $\qquad$ + $H_2O \rightarrow H_3O^+$ $\qquad\qquad$ + Cl$^-$

0.05 mol dm^{-3} $\qquad\qquad$ 0.05 mol dm^{-3}

So, the pH value is $-\log_{10}(0.05)$

$\qquad = 1.30$

Table 2.1 gives a series of values of the dissociation constants, molarities, hydronium ion concentrations, percentage dissociations and pH values for a selection of weak acids.

Table 2.1   Hydrogen ion concentration in weak acids

Acid	Dissociation constant	Molarity	$[H_3O^+]$	Percentage dissociation	pH
CH_3COOH	1.80×10^{-5}	0.20	1.91×10^{-3}	0.95	2.72
		0.10	1.35×10^{-3}	1.34	2.87
		0.05	9.58×10^{-4}	1.90	3.02
		0.01	4.33×10^{-4}	4.24	3.36
C_6H_5COOH	6.00×10^{-5}	0.20	3.49×10^{-3}	1.73	2.46
		0.10	2.48×10^{-3}	2.45	2.61
		0.05	1.76×10^{-3}	3.46	2.75
		0.01	8.05×10^{-4}	7.75	3.09
H_3BO_3	5.55×10^{-10}	0.20	1.05×10^{-5}	0.005	4.98
		0.10	7.45×10^{-6}	0.007	5.13
		0.05	5.27×10^{-6}	0.010	5.28
		0.01	2.36×10^{-6}	0.024	5.63
HF	5.62×10^{-4}	0.20	1.09×10^{-2}	5.30	1.96
		0.10	7.78×10^{-3}	7.50	2.11
		0.05	2.67×10^{-3}	10.60	2.25
		0.01	1.06×10^{-5}	23.71	2.57

$0.05\ M$ ammonia solution has a hydroxide concentration given by

$$[OH^-] = M\alpha$$

$$= 0.05 \times 1.87 \times 10^{-2}\ \text{mol dm}^{-3}$$

$$= 9.35 \times 10^{-4}$$

Since $[H_3O^+]\,[OH^-] = 10^{-14}$ in an aqueous system,

$$[H_3O^+] = 10^{-14}/9.35 \times 10^{-4}$$

$$= 1.07 \times 10^{-11}$$

$$pH = -\log_{10}[H_3O^+]$$

$$= 11.0$$

In general, acidic solutions (ones in which $[H_3O^+] > [OH^-]$) have pH values below 7 (Fig. 2.1).

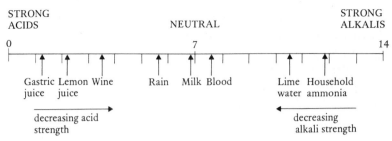

Fig. 2.1   The pH scale

It is useful to employ the terms pK_a and pK_b which have meanings similar to pH:

$$pH = -\log_{10} [H_3O^+]$$
$$pK_a = -\log_{10} K_a$$
$$pK_b = -\log_{10} K_b$$

Benzoic acid has, therefore, a pK_a value of $-\log_{10}$ (6.00×10^{-5}), i.e. 4.22. Ammonia solution has $pK_b = -\log_{10}$ (1.74×10^{-5}) = 4.76. pK_a and pK_b are related by the expression

$$pK_a = 14.0 - pK_b$$

So, for ammonia solution, the pK_a of the conjugate acid (NH_4^+) is given by

$$pK_a = 14.0 - 4.76$$
$$= 9.24$$

Strong acids have low pK_a values (approximately 1.0 or less); weak acids have larger values (e.g. boric acid is 9.2). Similarly, weak bases have high pK_b values (e.g. beryllium hydroxide has a value of 10.3); strong bases have low values. Whereas pH values vary with concentration (see Table 2.1), pK_a values are independent of concentration.

Buffer solutions

Water has a pH of 7. What is the pH of a solution prepared from 1 cm^3 1 M hydrochloric acid added to 1 dm^3 of water? The amount of acid added is 10^{-3} mole, so the concentration of the acidified solution is 10^{-3} mol dm^{-3}. Since the acid is a strong electrolyte, its pH = $-\log_{10} 10^{-3}$ = 3.0. So, the addition of 1 cm^3 of 1 M acid to 1 dm^3 of water causes its pH to drop from 7 to 3. In the body, the blood is of pH 7.4. If the pH goes outside the range 7.2–7.6, the result is fatal. There is obviously a need to keep the blood pH within these close limits. This is achieved by means of buffer solutions; these are, for example, solutions of weak acids and their salts.

A *buffer solution* is a solution whose pH is fairly constant even on the addition of small amounts of acid or alkali.

One buffer solution is prepared from sodium ethanoate in ethanoic acid solution. The sodium ethanoate dissociates into its constituent ions, and the ethanoic acid is partially dissociated into ions:

$$CH_3COONa(aq) \rightarrow CH_3COO^-(aq) + Na^+(aq)$$
$$CH_3COOH(aq) \rightleftharpoons CH_3COO^-(aq) + H^+(aq)$$

If acid is added, that is hydrogen ions, these combine with the ethanoate ions to produce more ethanoic acid molecules since this is a weak acid.

$$CH_3COO^-(aq) + H^+(aq) \rightarrow CH_3COOH(aq)$$

If hydroxide ions are added, these combine with hydrogen ions of ethanoic acid;

$$H^+(aq) + OH^-(aq) \rightleftharpoons H_2O$$

more acid molecules dissociate to replace the reacted hydrogen ions. Consider a mixture of 0.350 mole ethanoic acid and 0.650 mole sodium ethanoate in 1 dm^3 of solution.

$$K_a = \frac{[H^+]\ [CH_3COO^-]}{[CH_3COOH]} = 1.85 \times 10^{-5}$$

The ethanoate ion concentration is due to the sodium ethanoate (0.650 mole) and a small amount from the ethanoic acid. This latter amount is negligible compared to that from the sodium salt. So, $[CH_3COO^-] \approx 0.650\ M$. The ethanoic acid concentration is 0.350 M less a small amount due to the dissociation of the acid, so $[CH_3COOH] \approx 0.350\ M$.

$$1.85 \times 10^{-5} = [H^+] \times \frac{0.650}{0.350}$$

$$[H^+] = 1.85 \times 10^{-5} \times \frac{0.350}{0.650}$$

$$= 9.96 \times 10^{-6}$$

$$\therefore pH = 5.00$$

If we now add 1 cm^3 1 M hydrochloric acid, that is, 10^{-3} moles hydrogen ions, these will combine with ethanoate ions to form ethanoic acid molecules. So, the concentration of ethanoate ions is $(0.650 - 0.001) = 0.649\ M$. The concentration of ethanoic acid molecules is increased by the same amount; $(0.350 + 0.001) = 0.351\ M$ which is the new concentration of the undissociated acid.

$$So,\ 1.85 \times 10^{-5} = [H^+] \times \frac{0.649}{0.351}$$

$$[H^+] = 1.85 \times 10^{-5} \times \frac{0.351}{0.649}$$

$$= 1.00 \times 10^{-5}$$

$$\therefore pH = 5.00$$

The ethanoate solution has effectively buffered the solution at pH 5. In fact, it is possible to add up to 5 cm^3 1 M hydrochloric acid without any change in the pH value 5.00. The addition of 15 cm^3 of this acid has only a small effect (pH 5.01).

Each buffer solution has a characteristic known as its **buffer capacity**. This is the amount of acid or base that can be added to the solution without it suffering a change in pH. It is found that buffer solutions are not satisfactory if the pH of the solution lies outside of the range

$$pH = pK_a \pm 1$$

These ranges are given in Table 2.2 for some buffer mixtures. The exact pH of the solution is determined by the proportions of salt and acid (or base).

Table 2.2 Buffer solutions

pH range	Buffer mixture
3−6	$CH_3COOH - CH_3COONa$
6−8	$NaH_2PO_4 - Na_2HPO_4$
8−10	$Na_2B_4O_7 - NaOH$
11	$NH_4Cl - NH_3$

The blood-stream has to be buffered at pH 7.4 by means of carbon dioxide and hydrogen carbonate ions (HCO_3^-). The pK_a value for dissolved carbon dioxide in the blood is 6.1, so $K_a = 7.9 \times 10^{-7}$.

$$CO_2 + H_2O \rightleftharpoons H_2CO_3 \rightleftharpoons H^+(aq) + HCO_3^-(aq)$$

$$K_a = \frac{[H^+]\ [HCO_3^-]}{[H_2CO_3]}$$

If the pH = 7.4, then $[H^+] = 4.0 \times 10^{-8}$, so

$$\frac{[HCO_3^-]}{[H_2CO_3]} = \frac{7.9 \times 10^{-7}}{4.0 \times 10^{-8}} = \frac{19.8}{1}$$

So, to maintain a pH of 7.4, the ratio of the concentrations of HCO_3^- to H_2CO_3 must be 19.8 : 1.

Gaseous systems

Many industrial processes involve gas phase reactions. For example, the preparation of sulphuric acid (via sulphur trioxide),

$$2SO_2 + O_2 \rightleftharpoons 2SO_3$$

the synthesis of ammonia,

$$N_2 + 3H_2 \rightleftharpoons 2NH_3$$

and the oxidation of ammonia in the preparation of nitric acid

$$4NH_3 + 5O_2 \rightleftharpoons 4NO + 6H_2O$$

These systems can be treated in the same manner as the solutions already described. Consider the reaction between hydrogen and iodine vapour

$$H_2 (g) + I_2 (g) \rightleftharpoons 2HI(g)$$

We can write the equilibrium constant

$$K_c = \frac{[HI]_e^2}{[H_2]_e [I_2]_e}$$

and so calculate the value of K_c. However, it is normally more convenient to measure quantities of gases by their pressures rather than by their concentrations. The pressure of a gas is proportional to its molar concentration:

$$p \cdot V = n \cdot R \cdot T \quad \text{(general gas equation)}$$

$$\therefore p = \frac{n}{V} \cdot R \cdot T$$

$$= c \cdot R \cdot T$$

So, p is proportional to the molar concentration (c) at a constant temperature.

The equilibrium term can now be expressed as

$$K_p = \frac{P_{HI}^2}{P_{H_2} \cdot P_{I_2}} = \frac{[HI]^2 (RT)^2}{[H_2] (RT) [I_2] (RT)} = \frac{[HI]^2}{[H_2][I_2]} = K_c$$

where P_X is the equilibrium pressure of component X. The terms K_c and K_p are not necessarily equal in value (though in this case they are), but they are related. Since the pressure of a gas is related to the number of moles of that gas per unit of volume, the pressure of each component in a mixture is equal to the mole fraction multiplied by the total pressure. For example, if one-third of a gaseous mixture of total pressure 101 kPa is A, then the pressure of A is ($\frac{1}{3} \times 101$) kPa = 33.67 kPa.

In a reaction of equimolar quantities of hydrogen and iodine, it was found that at equilibrium there were 0.11 mole hydrogen, 0.11 mole of iodine and 0.78 mole hydrogen iodide. The total pressure at equilibrium was 600 kPa. What is the value of K_p?

Total number of moles = $(2 \times 0.11) + 0.78 = 1.00$.
Fraction of hydrogen in the mixture = $0.11/1.00 = 0.11$;
so, pressure of hydrogen = $0.11 \times 600 = 66.0$ kPa.
Similarly, the pressure of iodine = 66.0 kPa.
The pressure of hydrogen iodide = $0.78 \times 600 = 468.0$ kPa.
So,

$$K_p = \frac{P_{HI}^2}{P_{H_2} \cdot P_{I_2}} = \frac{468.0^2}{66.0 \times 66.0}$$

$$= 50.3$$

As with liquids, we can use the equilibrium constant to predict yields. For example, nitrogen and oxygen react to give nitrogen(II) oxide at high temperatures. If $K_p = 2.9 \times 10^{-2}$ at 2000 K for the reaction

$$\tfrac{1}{2} N_2\,(g) + \tfrac{1}{2} O_2\,(g) \rightleftharpoons NO(g),$$

what is the percentage of nitrogen(II) oxide formed when air (79% nitrogen, 21% oxygen by volume) is heated to this temperature?

For gases the number of moles is proportional to the volume. So, air contains 0.79 moles nitrogen to 0.21 moles oxygen. So, if the amount of nitrogen converted is x, then
the amount of nitrogen at equilibrium is $(0.79 - x)$,
the amount of oxygen at equilibrium is $(0.21 - x)$,
and the amount of nitrogen(II) oxide is $2x$. (Since the gases react in equimolar proportions, x moles of oxygen are consumed by x moles of nitrogen; $2x$ moles of nitrogen(II) oxide are formed.) The equilibrium pressures of the gases are

$$P_{N_2} = \frac{(0.79 - x) \cdot 101.3}{1.00}\ kPa$$

$$P_{O_2} = \frac{(0.21 - x) \cdot 101.3}{1.00}\ kPa$$

$$P_{NO} = \frac{2x \cdot 101.3}{1.00}\ kPa$$

where the total number of moles is 1.00 and the atmospheric pressure is 101.3 kPa.
So, for the equilibrium constant

$$K_p = \frac{P_{NO}}{P_{N_2}^{\frac{1}{2}} \cdot P_{O_2}^{\frac{1}{2}}}$$
$$= 2.9 \times 10^{-2}$$

we have, by substitution,

$$2.9 \times 10^{-2} = \frac{2x \cdot 101.3}{[(0.79 - x) \cdot 101.3]^{\frac{1}{2}} \cdot [(0.21 - x) \cdot 101.3]^{\frac{1}{2}}}$$

Hence,

$$x = 5.8 \times 10^{-3}$$

The fraction of nitrogen oxide in the mixture is, therefore,

$$\frac{5.8 \times 10^{-3} \times 2}{1.00}$$

that is, 1.16 per cent.

Heterogeneous systems

Systems involving substances in more than one physical phase are known as heterogeneous systems.

(a) Gas-solid equilibria

When calcium carbonate is heated in a closed vessel, an equilibrium is set up:

$$CaCO_3 (s) \rightleftharpoons CaO(s) + CO_2 (g)$$

Why must the container be closed? If the apparatus is open to the atmosphere, the carbon dioxide can escape and an equilibrium can not be established. The equilibrium is established in the gaseous phase only, so it is found that extra quantities of solid can be introduced without affecting the amount of carbon dioxide produced. The position of the equilibrium is, therefore, unaffected by the amount of solid present; it is determined solely by the pressure of the carbon dioxide. So,

$$K_p = P_{CO_2}$$

The pressure term must, of course, be the pressure at equilibrium. The extent of decomposition (*under equilibrium conditions*) is determined solely by the temperature (Fig. 2.2).

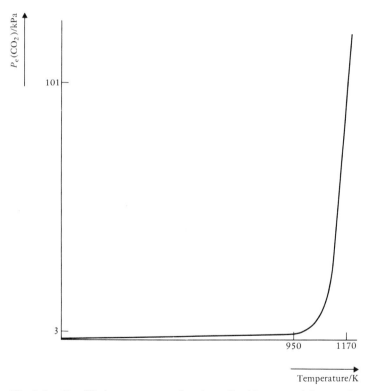

Fig. 2.2   Equilibrium pressure of carbon dioxide over calcium carbonate

Similar principles apply to the phenomena of efflorescence and deliquescence. **Efflorescence** is the process in which a hydrated solid loses water of crystallisation on exposure to the atmosphere. For example, sodium carbonate decahydrate is a colourless crystalline solid, which, on exposure to the atmosphere, loses some of its water of crystallisation to produce a white powder, the monohydrate:

$$Na_2CO_3 \cdot 10H_2O \rightleftharpoons Na_2CO_3 \cdot H_2O + 9H_2O$$

In a closed vessel equilibrium is established, preventing extensive loss of water. That is one good reason for keeping reagent bottles closed during a practical session!

The reverse process, the taking up of water, is known as **deliquescence.** Usually a distinction is drawn between **hygroscopic** substances which absorb water from the atmosphere and remain as dry solids (e.g. NaCl) and deliquescent materials which absorb sufficient moisture to form solutions (e.g. $CaCl_2 \cdot 6H_2O$, NaOH).

Figure 2.3 illustrates the equilibria that exist between hydrates of copper(II) sulphate. If the water vapour pressure in the atmosphere is less

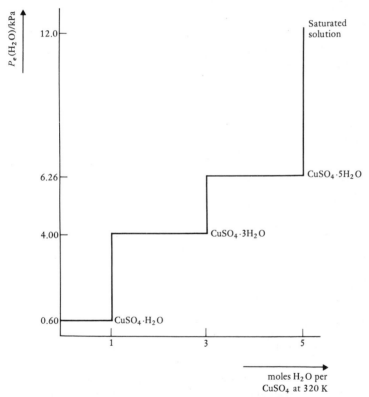

Fig. 2.3   Water vapour pressure and copper(II) sulphate

than 6.26 kPa at 320 K, then the pentahydrate will effloresce to give the trihydrate, because this is the form that exists in equilibrium with water vapour between 4.00 and 6.26 kPa. If the atmosphere is very humid, with water vapour pressure greater than 12.0 kPa, then the same substance is deliquescent and will form a saturated solution.

Some equilibria involve two or more gases in contact with solids. An example of this is the equilibrium between carbon, oxygen and carbon dioxide:

$$C(s) + O_2(g) \rightleftharpoons CO_2(g)$$

For this reaction the amount of excess solid has no effect on the equilibrium and so the constant, K_p, is given by

$$K_p = \frac{P_{CO_2}}{P_{O_2}}$$

Similarly, for the thermal dissociation of ammonium chloride,

$$NH_4Cl(s) \rightleftharpoons NH_3(g) + HCl(g)$$

the equilibrium constant is

$$K_p = P_{NH_3} \cdot P_{HCl}$$

Iron reacts with steam according to the equation

$$3Fe(s) + 4H_2O(g) \rightleftharpoons Fe_3O_4(s) + 4H_2(g)$$

This gives an equilibrium constant expression

$$K_p = \frac{P_{H_2}{}^4}{P_{H_2O}{}^4}$$

since 4 moles of each gas are involved in the stoichiometry of the reaction. It is possible to take the fourth root of the latter expression and obtain the simpler relationship

$$K_p' = \sqrt[4]{K_p} = \frac{P_{H_2}}{P_{H_2O}}$$

(b) Solid–solution equilibria

Excess solids also exist in contact with solutions. These are normally described as saturated solutions. For example, silver chloride is not very soluble in water; its solubility is 1.25×10^{-5} M at 298 K. So, the addition of 0.1 mole silver chloride to 1 dm^3 water produces an equilibrium:

$$AgCl(s) \rightleftharpoons Ag^+(aq) + Cl^-(aq)$$

(Evidence for the dynamic nature of this system is given in *Fundamentals of Chemistry*, p. 9.) The amount of solid present in excess has no measurable effect on the ionic concentrations, and so the position of equilibrium (and the equilibrium constant) is independent of the [AgCl(s)] term. So,

$$K_s = [Ag^+(aq)]_e \cdot [Cl^-(aq)]_e$$

The constant in these situations is described as the **solubility product,** K_s. In a saturated solution, the product of the molar ionic concentrations cannot exceed the value of the solubility product at that temperature. This is true only for solutions of sparingly soluble salts.

Lead(II) iodide has the equilibrium

$$PbI_2 (s) \rightleftharpoons Pb^{2+}(aq) + 2I^-(aq)$$

and the equilibrium expression

$$K_s = [Pb^{2+}(aq)]_e \cdot [I^-(aq)]_e^2$$

Note the squared value of the iodide concentration, which is due to the value of two for its molecularity in the equation.

It is possible to relate the solubility (in mol dm^{-3} or g dm^{-3}) to the solubility product. Silver chloride is often described as 'insoluble in water'. It is customary to classify substances as insoluble in a solvent if they do not form a solution of at least 1.0×10^{-3} mol dm^{-3} concentration. However, silver chloride does have a small solubility, 1.25×10^{-5} mol dm^{-3} in water. So, in 1 dm^3 water, 1.25×10^{-5} moles of AgCl will dissolve. It is a strong electrolyte and so is completely dissociated into its ions. 1.25×10^{-5} moles of AgCl will give 1.25×10^{-5} moles of both Ag$^+$ and Cl$^-$ ions. The molar concentration of each ion is, therefore, 1.25×10^{-5} M.

$$
\begin{aligned}
K_s &= [Ag^+] [Cl^-] \\
&= (1.25 \times 10^{-5})(1.25 \times 10^{-5}) \text{ mol}^2 \text{ dm}^{-6} \\
&= 1.56 \times 10^{-10} \text{ mol dm}^{-6}
\end{aligned}
$$

Lead(II) iodide has a solubility of 1.65×10^{-3} mol dm^{-3}. But 1.65×10^{-3} moles of lead(II) iodide will give 1.65×10^{-3} moles Pb^{2+} ions and $2 \times (1.65 \times 10^{-3})$ moles of iodide ions (2 moles iodide ions and 1 mole lead ions are produced per mole of lead iodide dissolved). Therefore,

$$
\begin{aligned}
K_s &= [Pb^{2+}(aq)] [I^-(aq)]^2 \\
&= (1.65 \times 10^{-3})(3.10 \times 10^{-3})^2 \text{ mol}^3 \text{ dm}^{-9} \\
&= 1.59 \times 10^{-8} \text{ mol}^3 \text{ dm}^{-9}
\end{aligned}
$$

The solubility product of barium sulphate (BaSO$_4$) at 298 K is 1.08×10^{-10} mol^2 dm^{-6}. What is the solubility of the salt at this temperature? The solubility product for this substance is of the form

$$K_s = [Ba^{2+}(aq)] [SO_4^{2-}(aq)]$$

So,

$$1.08 \times 10^{-10} = [Ba^{2+}(aq)] [SO_4^{2-}(aq)]$$

Since these ions are of equal concentration (1 mole of each per mole of dissolved barium sulphate),

$$
\begin{aligned}
[Ba^{2+}(aq)] &= (1.08 \times 10^{-10})^{\frac{1}{2}} \text{ mol dm}^{-3} \\
&= 1.04 \times 10^{-5} \text{ mol dm}^{-3}
\end{aligned}
$$

This concentration of ions is produced by the same concentration of dissolved barium sulphate (1 mole of Ba^{2+} from 1 mole $BaSO_4$), so

$$[BaSO_4(aq)] = 1.04 \times 10^{-5} M$$

If it is desirable to quote the concentration as a mass per dm^3, then the concentration of $BaSO_4$ in a saturated solution is

$$(1.04 \times 10^{-5}) \times 233.4 \text{ g dm}^{-3} = 2.43 \times 10^{-3} \text{ g dm}^{-3}$$

The solubility product can be used to determine the solubility of the salt in solvents other than pure water. For example, what is the solubility of barium sulphate in $0.01 M$ sulphuric acid?

$$K_s = [Ba^{2+}(aq)] \ [SO_4^{2-}(aq)] = 1.08 \times 10^{-10}$$

In this case, while Ba^{2+} ions are derived solely from the salt, the SO_4^{2-} ions are present due to the salt and the acid. Since the acid is strong, the sulphate ion concentration is mainly due to the acid and is 0.01 mol dm^{-3}. This is much larger than the concentration due to the salt and so the latter contribution can be considered as insignificant.

So,
$$[SO_4^{2-}(aq)] = 0.01 \text{ mol dm}^{-3}$$
$$1.08 \times 10^{-10} = [Ba^{2+}(aq)] \cdot (0.01)$$
$$[Ba^{2+}(aq)] = 1.08 \times 10^{-10}/0.01$$
$$= 1.08 \times 10^{-8} \text{ mol dm}^{-3}$$

Therefore, the concentration of dissolved barium sulphate is 1.08×10^{-8} mol dm^{-3}. It will be observed that this is smaller than the amount dissolving in water by a factor of 10^3. This can be of importance in the washing of precipitates in which a quantitative separation is required. If a precipitate of barium sulphate (0.1 g) is obtained and it is washed with water (100 cm^3), there will be a loss of $2.43 \times 10^{-3} \times 0.1$ g, 100 cm^3 being 0.1 dm^3. So, the residual mass of precipitate will be $(0.100\ 0 - 0.000\ 2) = 0.099\ 8$ g. This represents a 0.2 per cent loss of material. (The percentage loss is, of course, smaller for larger amounts of precipitate, and is more significant for smaller quantities of solid.) If $0.01 M$ sulphuric acid (100 cm^3) is used, the mass loss is 2.52×10^{-7} g – well below the level detectable on conventional analytical balances.

(c) Liquid-liquid systems

If a solute X is added to two immiscible liquids, A and B, an equilibrium is set up, provided X is soluble in both solvents:

X in A $\rightleftharpoons$ X in B,

and the equilibrium description is

$$K = \frac{\text{Concentration of X in B}}{\text{Concentration of X in A}}$$

The term, K, is known as the partition coefficient. This mathematical expression is formulated in the 'Partition law', which states that 'if a solute X has the same molecular state in two immiscible solvents A and B, it will distribute itself between them so that the concentrations of X in A and of X in B are in a constant ratio'.

Iodine dissolves in both water and toluene. If the two solvents are mixed they separate into two layers — toluene floating on the water. If iodine is added to the two layers it dissolves in both. In a typical experiment, 1 g iodine is shaken with 50 cm^3 toluene and 250 cm^3 water in a stoppered flask for 30 minutes. Portions of each layer are removed and analysed for iodine. A further 0.5 g iodine is added. After shaking again, portions are analysed for iodine. After the addition of a further portion of iodine, the process is repeated. Table 2.3 gives typical results of these analyses. We see from this table that the ratio of the concentrations is effectively constant.

Table 2.3   Concentration of iodine in a toluene—water mixture

Portion	Concentration of iodine in water (C_w) / g dm^{-3}	Concentration of iodine in toluene (C_t) / g dm^{-3}	C_t/C_w
1	0.058	2.973	51.26
2	0.087	4.457	51.23
3	0.115	5.945	51.70

In the example cited above, the average partition coefficient of iodine between toluene and water is 51.39. The value is dependent on the nature of the substances concerned and the temperature. It is independent of the amounts of the solvents. If solvent A is saturated with X, then so is B.

Some solutes have differing forms in the separate solvents. For example, ethanoic acid is in the associated form $(CH_3COOH)_2$ in benzene. Table 2.4 gives the results of an experiment using benzoic acid distributed between water and benzene. Benzoic acid has a similar associated form. It is found that when n units of the solute associate to form a complex structure in the solvent, the nth root of the concentration of the solute in that solvent must be used; $n = 2$ in Table 2.4.

Table 2.4   Distribution of benzoic acid between benzene and water

Portion	Concentration of benzoic acid in water (C_w) / g dm^{-3}	Concentration of benzoic acid in benzene (C_b) / g dm^{-3}	$\dfrac{C_b}{C_w}$	$\dfrac{\sqrt{C_b}}{C_w}$
1	0.015 0	0.242	16.13	32.80
2	0.019 5	0.412	21.13	32.92
3	0.028 9	0.870	30.10	32.27

The partition law is modified to state that 'for the distribution of a solute X between two immiscible solvents A and B, X being associated into X_n in A but unassociated in B, then the concentration of X in B to the nth root of the concentration of X in A is constant'.

This principle of partition is applied in solvent extraction (see p. 103).

Free energy and equilibrium

In *Fundamentals of Chemistry* (p. 156), it was shown that the e.m.f. of a cell (the potential under which the cell will do the maximum amount of work)

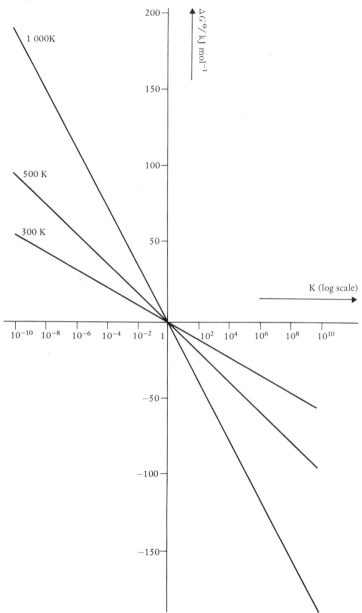

Fig. 2.4   Variation in free energy change with the equilibrium constant at different temperatures

can be expressed as an energy equivalent, the free energy change (ΔG)

$$\Delta G = -n \cdot E \cdot F$$

where E is the e.m.f. and F is the Faraday.

Since the e.m.f. is dependent on the concentrations of the participating species (*Fundamentals of Chemistry*, p. 147), it follows that ΔG too is affected by the concentrations. It can be shown that for an equilibrium system, the standard free energy change ($\Delta G^{\ominus}$) is given by

$$\Delta G^{\ominus} = -R \cdot T \cdot \log_e K$$

where K is the equilibrium constant. Figure 2.4 shows the variation in $\Delta G^{\ominus}$ with K at three different temperatures.

The standard e.m.f. of the reaction

$$Cu^{2+}(aq) + Zn(s) \rightleftharpoons Zn^{2+}(aq) + Cu(s)$$

is 1.107 V. So, the $\Delta G^{\ominus}$ value is $-n \cdot E \cdot F$

$$= -2 \times 1.107 \times 96\ 487\ J$$
$$= -2.136 \times 10^5\ J$$

Therefore, $\log_e K = (-2.136 \times 10^5)\ /\ (-8.314 \times 300)$

$$= 85.64$$

So, $K = 1.53 \times 10^{37}$

(This same value can be obtained by extrapolation of the 300 K line in Fig. 2.4.)

Free energy and enthalpy

It was seen in *Fundamentals of Chemistry* (p. 162), that the change in internal energy, when there is no change in volume or pressure, is equal to the enthalpy change, ΔH. For the zinc−copper reaction described above, the enthalpy change is 209.8 kJ mol^{-1}. This deviates slightly from the value for the free energy change, calculated above as 213.7 kJ mol^{-1}. Careful experimentation has shown that this difference is not due to experimental errors. In fact, in many reactions the deviation is larger. The gap between these values becomes more significant as larger temperature ranges are considered. It becomes apparent, therefore, that the maximum work available from a cell (the e.m.f.) is not equal to the change in internal energy even when no gas is evolved.

The deviation, $\Delta H - \Delta G$, is dependent on temperature and another factor, ΔS, the entropy change.

$$\Delta H - \Delta G = T \cdot \Delta S$$

This is usually rearranged to give

$$\Delta G = \Delta H - T \cdot \Delta S$$

since it is the free energy change that determines the viability of a reaction. As can be seen from Fig. 2.4, a negative value for ΔG gives an equilibrium constant greater than unity, so the equilibrium lies towards the products.

The term $T\Delta S$ is a measure of the availability of the energy for the performance of useful work. If there is an increase in the number of molecules as a result of a reaction, the energy is shared more widely; it is, therefore, more difficult to utilise it. This can be illustrated by an analogy. If a man lent £100 to one hundred people at random, he would have more difficulty in recovering the money than if he lent it to a single friend. Similarly, the money could be more readily recovered from a man in a small room than from one in a concert hall, since he would be easier to catch!

In a reaction in which a gas is produced, the term $T \cdot \Delta S$ is relatively large and causes a significant deviation between ΔG and ΔH. In these cases, the energy is distributed between the gaseous molecules which are moving randomly about a volume larger than that of the original substances.

Consider the reaction

$$CuCl_2 \cdot 2H_2O(s) \rightleftharpoons CuCl_2 \cdot H_2O(s) + H_2O(g)$$

The relevant values at 353 K are $\Delta G^{\ominus} = +3.33$ kJ mol^{-1}

$$\text{and } \Delta H^{\ominus} = +57.29 \text{ kJ mol}^{-1}$$

There is a substantial difference between these values (contrast the difference with that between those given above for the zinc–copper reaction). This gives a value of 0.15 kJ mol^{-1} for the entropy change in this reaction. The value given for the zinc–copper reaction is -0.01 kJ mol^{-1} at 293 K. Since the ΔS value does not change substantially with temperature change, this value will be approximately correct at 353 K as well.

A positive entropy change (as in the first example) indicates that the energy has been more widely dissipated than at the beginning of the reaction, as in reactions involving an increase in the number of molecules or in the volume occupied by the system. Conversely a negative entropy change corresponds to an increase of order in the system.

Reactions that occur spontaneously have negative free energy changes and usually positive entropy changes as well.

Summary

At the conclusion of this chapter, you should be able to:

1. state that an acid is a substance that donates hydrogen ions in a reaction;
2. state that a base is a substance which accepts hydrogen ions in a reaction;
3. identify acids and bases in chemical equations;
4. describe the meaning of the terms conjugate acid and conjugate base;
5. distinguish between a strong and a weak acid or base;
6. write the expression for the dissociation constant of an acid, HA(aq);
7. derive the relationship between the degree of dissociation, α, and concentration, M mol dm^{-3}, for a dilute solution of a weak acid;
8. calculate the degree of dissociation for a weak acid, given the molarity and dissociation constant;

9. define the meaning, and state the value, of the ionic product of water;
10. define the term pH value;
11. calculate the pH value of an acid or base;
12. define the terms pK_a and pK_b;
13. define the terms buffer solution and buffer capacity;
14. explain the effectiveness of a buffer solution;
15. write the equilibrium expression for a gaseous reaction in terms of partial pressures;
16. calculate the yield in a gaseous reaction, given the equilibrium constant;
17. write the equilibrium constant for a heterogeneous reaction involving gases and solids;
18. explain the phenomena efflorescence and deliquescence;
19. define the term solubility product;
20. calculate the value of the solubility product, given the solubility of a sparingly soluble salt;
21. explain the significance of solubility product to the washing of precipitates;
22. state the partition law;
23. determine the value of the partition coefficient, given the concentrations of solute in two immiscible solvents;
24. relate the free energy change to the equilibrium constant;
25. relate the free energy change to the enthalpy change, entropy change and temperature of a reaction.

Experiments

2.1 Variation in pH with concentration

Determine the pH of a 0.1 M solution of hydrochloric acid using a pH meter. Transfer 1 cm^3 of the acid to a measuring cylinder and dilute to 10 cm^3. Determine the pH of this 0.01 M solution.

Repeat the determination on the following solutions:

(a) 1 cm^3 0.1 M acid diluted to 100 cm^3, giving a 0.001 M solution;
(b) 1 cm^3 0.01 M acid diluted to 100 cm^3, giving a 0.000 1 M solution;
(c) 1 cm^3 0.001 M acid diluted to 100 cm^3, giving a 0.000 01 M solution.

Further dilutions may be obtained if required. Plot the pH value against the molarity of the acid.

The procedure may be repeated for a weak acid (e.g. 0.1 M ethanoic acid) and for bases.

2.2 Properties of buffer solutions

(a) Place equal volumes of 2 M ammonia solutions (5 cm^3) in two test tubes. To one add an equal volume of deionised water and to the other add the same volume of 2 M ammonium chloride solution. Measure the pH of each solution.

(b) Place equal volumes of 2 M ethanoic acid solution (5 cm^3) in two test tubes. To one add water (5 cm^3) and to the other add the same volume of

2 M sodium ethanoate solution. Measure the pH of each solution.
(c) Place 5 cm^3 deionised water in one test tube and 5 cm^3 0.2 M sodium ethanoate in another test tube. To each add 2 drops of methyl red indicator. To the first tube add 1 cm^3 0.01 M hydrochloric acid. Add the same acid to the second tube until the two test tubes have the same indicator colour. How much acid was required in the second case?

Account for the observations in these experiments.

2.3 Preparation and properties of buffer solutions

Buffer solutions in the range pH 4.0−9.0 should be prepared as shown in the table below from 0.1 M solutions of sodium dihydrogen phosphate and disodium hydrogen phosphate.

Volume of 0.1 M NaH$_2$PO$_4$(aq)/cm^3	10.0	8.0	4.0	2.0	0.0
Volume of 0.1 M Na$_2$HPO$_4$(aq)/cm^3	0.0	2.0	6.0	8.0	10.0
pH value	4.0	6.0	6.8	7.2	9.0

To 1 cm^3 portions of each mixture, add 1 cm^3 water and 1 drop of universal indicator. Compare the efficiency of these solutions in buffering against acid, by adding 0.01 M hydrochloric acid dropwise and noting the pH change against the number of drops added.

Determine the buffering effect of the solutions against 0.01 M sodium hydroxide solution.

2.4 Determination of the partition coefficient of ethanoic acid between water and tetrachloromethane

Fill five stoppered flasks with the mixtures indicated in the table below and shake thoroughly. Allow to stand, with occasional shakings, for a further 30 minutes.

Flask	1	2	3	4	5
1 M ethanoic acid/cm^3	5	10	15	20	25
Distilled water/cm^3	20	15	10	5	0
Tetrachloromethane/cm^3	25	25	25	25	25

Withdraw portions of the aqueous layer (10 cm^3) and titrate against 0.5 M sodium hydroxide solution.

Withdraw portions of the organic layer (10 cm^3) and titrate against 0.005 M sodium hydroxide solution.

Deduce the molecular state of the acid in the organic layer, assuming that it exists as CH$_3$COOH in the aqueous medium.

2.5 Determination of the solubility product of silver ethanoate at room temperature

Using the quantities indicated in the table below, stir mixtures of 0.2 M silver nitrate and 0.2 M sodium ethanoate solutions until precipitation is complete.

Filter off the precipitate and titrate portions of the filtrate (25 cm^3), to which has been added dilute nitric acid (20 cm^3), against $0.2 M$ ammonium thiocyanate solution using concentrated iron(III) alum solution (1 cm^3) as indicator (to a red end point).

Flask	1	2	3	4
Volume of $AgNO_3/\text{cm}^3$	50	40	30	20
Volume of $NaOOC \cdot CH_3/\text{cm}^3$	30	40	50	60

Calculate the silver ion and ethanoate ion concentrations in each case and so the solubility product. (The titration gives the silver ion concentration at equilibrium; a comparison with the initial silver ion concentration indicates the amount reacted; the amount of silver ion reacted will be equal to the amount of ethanoate ion reacted; hence calculate the ethanoate ion concentration in solution at equilbrium.)

References

An Introduction to Chemical Energetics; J. J. Thompson (Longman, 1967).
Energy Changes in Chemistry; J. A. Allen (Blackie, 1975).
pH values (BDH, 1977).

Films

Equilibrium (CHEM Study).
The laws of disorder (ICI).

Questions

1. Identify the acid and the base in the following chemical reactions:

    (a) $HCl + CH_3COOH \rightleftharpoons CH_3COOH_2^+ + Cl^-$

    (b) $CH_3COOH + NH_3 \rightleftharpoons CH_3COO^- + NH_4^+$

    (c) $H_2S + CO_3^{2-} \rightleftharpoons HS^- + HCO_3^-$

    (d) $H_2O + NH_3 \rightleftharpoons NH_4^+ + OH^-$

    (e) $NH_2^- + NH_4^+ \rightleftharpoons 2 NH_3$

2. Show that, for a dilute solution of a weak acid, the degree of dissociation is inversely proportional to the square root of its molarity. What is the degree of dissociation of a $0.1 M$ solution of methanoic acid (dissociation constant, 2.1×10^{-4})?

3. An acid HA has a dissociation constant K, a concentration M mol dm^{-3}, and a degree of dissociation α. Write an expression for its pH. Calculate the pH of a $0.01 M$ ethanoic acid solution for which $K = 1.8 \times 10^{-5}$.

4. Calculate the pH of a solution of 5.00 g NH_4Cl in 1 dm^3 of 1 M ammonia solution (K for ammonia solution = 1.7×10^{-5}). What is the pH after adding 1 cm^3 0.1 M solution of hydrochloric acid? ($K_w = 10^{-14}$).

5. The table below gives the vapour pressures of some solid hydrates and saturated solutions at 293 K. At this temperature the atmospheric water vapour pressure is 1.87 kPa. Predict whether the solid substance will be stable, efflorescent or deliquescent.

Substance	Water vapour pressure of hydrate/kPa	Vapour pressure of saturated solution/ kPa
$CuSO_4 \cdot 5H_2O$	0.67	2.13
$CaCl_2 \cdot 6H_2O$	0.33	1.00
$Na_2SO_4 \cdot 10H_2O$	2.17	2.21
$Na_2CO_3 \cdot 10H_2O$	3.23	3.47

6. A saturated solution of lead(II) chloride was filtered to remove the excess solid. The solution was found to contain 4.00×10^{-2} mol $PbCl_2$ per dm^3 of solution. Calculate the solubility product of lead(II) chloride. What would be the concentration of $PbCl_2$ in solution, if 0.5 mole KCl were added to 1 dm^3 of the solution?

7. The following equilibrium exists at 300 K:

$2NO_2 (g) \rightleftharpoons N_2O_4$

At a pressure of 100 kPa, there is 20 per cent NO_2 present. Calculate K_p for the reaction.

8. Copper(II) chloride dihydrate exists in equilibrium with water vapour as shown by the equation

$CuCl_2 \cdot 2H_2O (s) \rightleftharpoons CuCl_2 \cdot H_2O(s) + H_2O(g)$

The equilibrium water vapour pressure at 290 K is 1.2×10^{-5} kPa. Calculate the equilibrium constant, K_p, at this temperature. Hence determine the free energy change for this reaction.

9. The following results were obtained in an experimental investigation of the ratio of the distribution of phenol between water and trichloromethane. If the solute undergoes no dissociation in aqueous solution, what conclusions may be drawn regarding its molecular state in the trichloromethane solvent?

Concentration in aqueous solution (g dm^{-3})	0.94	1.63	2.47	4.36
Concentration in trichloromethane (g dm^{-3})	2.54	7.61	18.5	54.3

10. Briefly indicate how you would show experimentally that, for the reaction

$Zn(s) + Cu^{2+}(aq) \rightleftharpoons Zn^{2+}(aq) + Cu(s)$

the enthalpy change is -201.4 kJ mol^{-1} and that the standard e.m.f. is $+1.10$ volts (using 1 M solutions).

Calculate: (a) the free energy change; (b) the equilibrium constant; and (c) the entropy change for this reaction. Assume that all measurements were made at 27 $^{\circ}$C.

(F = 96 490 coulombs mole^{-1}; R = 8.314 J deg^{-1} mol^{-1})

Chapter 3

Kinetics

In Chapter 2 (p. 39) it was shown, from a knowledge of the equilibrium constant (or the free energy, p. 46), that it is possible to predict the yield of a reaction. It is possible, therefore, to ascertain whether or not a reaction is likely to yield the desired products. But the energetics do not tell us how fast the products are formed. This is a separate study known as the kinetics or the rate of a reaction.

The rate of a reaction is the rate at which a product is formed or a reactant is consumed. It is measured, for example, as the rate at which the

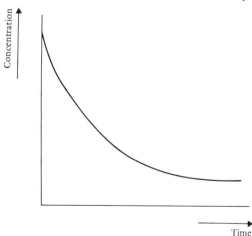

Fig. 3.1   Progress in a chemical reaction

concentration of a reactant decreases. In the reaction between permanganate and ethan-1, 2-dioate ions (experiment 3.1), the change in concentration of permanganate ions with time follows the pattern shown in Fig. 3.1. This is a typical graph for the progress of a chemical reaction, the gradient being determined by the nature of the chemicals and other factors (see below). The gradient is the rate of reaction.

This can be understood by use of an analogy. If a car is travelling along a road and a passenger notes the progress of the vehicle towards its destination by recording the time at a series of milestones, a graph such as Fig. 3.2

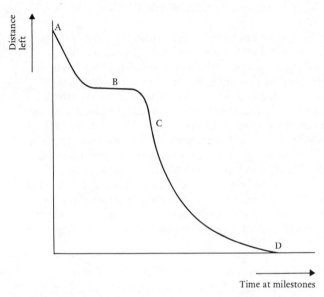

Fig. 3.2   Progress of a car

might be obtained. In this diagram, point A indicates the mileage to be covered at the start of the journey. There is a steady decrease in this distance until plateau B is reached. B indicates that no distance is covered for a period of time; that is, the car has stopped.

There is then a steep part of the curve, C, corresponding to a fast movement of the car; it covers a large distance in a small time. Gradually the curve becomes less steep, indicating a slowing down, until the destination D is reached (no distance left to travel). A steady linear decrease indicates a steady speed, a horizontal line corresponds to no movement, and the gradient is a measure of the actual speed. If a curve is obtained, rather than a straight line, then the car is either accelerating or decelerating.

In Fig. 3.1 we see that the amount of permanganate decreases with time (it is consumed), and that the rate of consumption gradually changes – the reaction slows down.

Factors affecting the reaction rate

(a) Concentration

The rate of a chemical reaction depends on the concentration of the reactants. As the concentration decreases, the rate of reaction decreases. Experiment 3.2 illustrates the same principle for another reaction – the reaction between peroxodisulphate and iodide ions. Notice that the time taken to produce the iodine–starch colour depends on the concentration of the initial reactants; halving the concentration of one reactant approximately doubles the time taken in this case. (The magnitude of the effect varies according to the reaction.) In general, an increase in the concentration of the reactants accelerates the reaction.

The concentration effect can be attained by other related factors. An increase in pressure in a gaseous reaction is analogous to an increase in concentration, and so this also affects the reaction rate. Pressure has a negligible effect on reactions in solution because the solvent is of low compressibility and the volume (and so the concentration) is not affected.

When a solid is used (for example, experiment 3.3) the rate is affected by particle size. This is because identical masses of solid of different particle size have differing surface areas. Different surface areas give variations in the number of available reaction sites. For example, a cube of edge 1 cm will have a total surface area of 6 cm^2. If the cube is divided into 8 cubes of 0.5 cm edge, the total area is doubled to 12 cm^2. Division into 1 000 cubes of edge 0.1 cm gives a total area of 60 cm^2. So, more atoms of the solid are exposed in the smaller particles than in the larger ones. Therefore, more atoms (e.g. of zinc in experiment 3.3) are available for reaction in a powder than in a sheet of foil.

(b) Temperature

The peroxodisulphate–iodide experiment also illustrates the effect of temperature. A decrease in temperature results in a slowing down of the reaction; an increase in temperature causes an acceleration of the change. It is often stated that an increase of 10° in the temperature doubles or trebles the rate of a chemical reaction. While this statement is an over-simplification, it is, nevertheless, a useful 'rule-of-thumb'.

(c) Catalysis

It is found that certain substances are able to increase the rate of a reaction without affecting the course of the reaction or being themselves consumed in the reaction. As will be seen later in the chapter, they do participate in the reaction in order to accelerate it, but are continually regenerated in the reaction. As a result, usually only small quantities are needed to be effective. In experiment 3.3, which ions have a catalytic effect on the reaction? These can be identified easily from the much lower reaction times obtained using these ions. In experiment 3.1, a small amount of manganese(II) ions were added to catalyse the reaction. Rerun the reaction as described in experiment 3.4. Figure 3.3 gives some typical experimental results for this

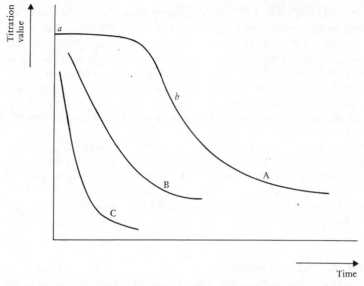

Fig. 3.3 Progress in a permanganate–oxalate reaction

experiment. Curve A in this figure differs from the general curve (Fig. 3.1) in that initially there is little reaction and then the gradient slowly increases. At the point b, the curve assumes the general form. What, therefore, is happening in the section a–b? The increasing gradient indicates that the reaction rate is increasing. Consider the chemical change occurring:

$$16H^+ + 2MnO_4{}^- + 5C_2O_4{}^{2-} \rightarrow 2Mn^{2+} + 5CO_2 + 8H_2O$$

No manganese(II) ions were present to catalyse the reaction in this run, so the reaction is very slow initially (hence the almost horizontal section). But, once the reaction has started, some of these ions are produced and so the products catalyse the reaction. This phenomenon of a reaction being accelerated by its own products is known as **autocatalysis.** Curves B and C do not show autocatalysis because the Mn^{2+} ions (in B) and the higher temperature (in C) increase the rate sufficiently to mask the effect.

Experimental methods

Any property that varies during a chemical change can, in principle, be used to follow the progress of a reaction. Some of the common techniques are listed.

(a) Absorbance (experiment 3.1) – if there is a change in the concentration of a coloured species, then the intensity of the transmitted colour, or the absorbance of incident light, can be used to follow the reaction.

(b) Gas volume (experiment 3.3) – if a gas is evolved in a reaction (or, in general, there is a change in the volume of a gas), then the rate of reaction is indicated by the rate of evolution of the gas.

(c) Titrimetry (experiment 3.4) – if the concentration of a reactant, or of a product, can be determined by titration, then the reaction rate can be followed by titration of the reaction mixture after specific time intervals.

(d) Conductance (experiment 3.5) – a change in the concentration (or nature) of the conducting particles (see *Fundamentals of Chemistry*, Ch. 12) provides a further method for following reactions.

(e) Precipitation (experiment 3.6) – the rate of formation of a precipitate may be followed by determining the opacity of a solution or the mass of precipitate formed.

(f) Specific rotation of the plane of polarised light – optically active sucrose (dextrorotatory) is hydrolysed to glucose and fructose (with a net laevorotatory aspect); as the concentration of the products increases, the rotation changes.

(g) Viscosity – when polymers are produced, the viscosity of a solution increases.

Theories of reaction rates

According to the **collision theory**, the reaction rate is determined by the number of effective collisions between reaction molecules in a second. For two molecules to react, they must collide with sufficient energy (known as the activation energy) to undergo a chemical change. If the molecules have insufficient energy to react, they separate unchanged. When the concentrations of the reacting molecules are increased, there is a better chance of collision, so the rate is increased. When the temperature is raised, the energy of the molecules is greater and so more collisions occur with an energy greater than the activation energy.

The rate of reaction, based on this approach, is expressed as

$$\text{rate} = Z \cdot e^{-E/RT}$$

where Z = the number of collisions per unit of time per unit of volume,
$\quad\quad E$ = the activation energy of the reaction,
$\quad\quad T$ = the temperature,

and the exponential term indicates the fraction of the collisions that are effective in causing a reaction.

It is found, however, that not every collision with the required energy is successful in bringing about a chemical change. This is because of a steric factor, P.

$$\text{rate} = P \cdot Z \cdot e^{-E/RT}$$

The fraction P indicates that only a portion of the molecules collide with the correct geometry. This may be illustrated by the reaction between hydrogen and iodine gas, as shown in Fig. 3.4. To produce the hydrogen iodide, the reactants must collide with the geometry shown. In this particular reaction, the collision theory predicts a reaction rate in good agreement with the experimental value.

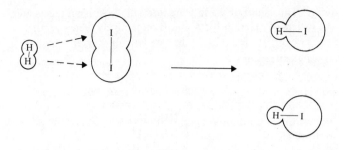

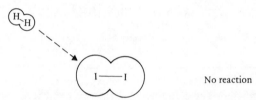

No reaction

Fig. 3.4 Collision geometry in the hydrogen–iodine reaction

This does not, of course, prove that the mechanism is correct, merely that it serves as a useful model here. Alternative models might be proposed, some of which also make successful predictions. In fact, recent experimental work seems to suggest that even in this hydrogen–iodine reaction this proposed mechanism is invalid.

It is found that this mechanism does not predict rates consistent with the experimental data if:

(*a*) heterogeneous catalysis is involved;

(*b*) a series of reactions occurs rather than one single step;

(*c*) large molecules, with complex stereochemical factors, are involved.

Some of these effects are considered again later in the chapter.

Order of reaction

It has been shown (p. 56) that the rate of a reaction is proportional to the concentration of the reacting species. In 1879, Guldberg and Waage showed that the reaction rate is proportional to the molar concentrations of the constituents, provided dilute solutions are used. This is the *law of mass action*. Expressed mathematically, for a reaction

$A + B \rightarrow$ products

rate $= k \cdot [A] \cdot [B]$, where the square brackets represent molar concentrations.

However, further study has shown that for many reactions an expression of the form

rate $= k \cdot [A]^x \cdot [B]^y$

is more appropriate. The values of x and y are normally positive integers and are known as the **orders of reaction** for A and B respectively. The overall order of reaction is $(x+y)$. Table 3.1 gives the rate law for several reactions.

Table 3.1 Rate law expressions

Reaction	Rate law expression
$H_2 + I_2 \rightarrow 2HI$	$k [H_2] [I_2]$
$2N_2O_5 \rightarrow 2N_2O_4 + O_2$	$k [N_2O_5]$
$2NO + O_2 \rightarrow 2NO_2$	$k [NO]^2 [O_2]$
$S_2O_8{}^{2-}(aq) + 2I^-(aq)$ $\rightarrow 2SO_4{}^{2-}(aq) + I_2 (aq)$	$k [S_2O_8{}^{2-}] [I^-]$
$BrO_3{}^-(aq) + 5Br^-(aq) + 6H^+(aq)$ $\rightarrow 3H_2O + 3Br_2 (aq)$	$k [BrO_3{}^-] [Br^-] [H^+]^2$
$4HBr + O_2 \rightarrow 2H_2O + 2Br_2$	$k [HBr] [O_2]$
$CH_3CHO \rightarrow CH_4 + CO$	$k [CH_3CHO]^{3/2}$
$2H_2O_2 \rightarrow 2H_2O + O_2$	$k [H_2O_2]$

It must be emphasised that the orders of reaction are experimentally determined and cannot be predicted. The constant, k, in the rate expressions is known as the **velocity** (or **rate**) **constant.** It is dependent on the reaction, the presence or absence of a catalyst, and the temperature. Experiment 3.7 illustrates a method by which the order of reaction may be determined. The reaction considered is that between hydrogen iodide and hydrogen peroxide. It is difficult to determine the order of reaction for both components simultaneously. The principle is to follow the variation in the concentration of one reactant (in this case, H_2O_2) and to keep the concentration of the other reactant constant. At first consideration this might appear impossible, since both reactants are consumed as the reaction proceeds. However, this problem can be overcome either: (*a*) by regenerating the required reagent (HI) as it is consumed; or (*b*) by using it in very large excess so that the variation in concentration is negligible. Situation (*a*) can be effected by adding thiosulphate ions, which reduce iodine to iodide ions. Unfortunately it then becomes difficult to follow the rate of reaction, for example, by titration with permanganate, since the thiosulphate also reacts with the titrant. The second method is used more frequently. In fact, this reaction is found to be dependent on the concentrations of H_2O_2, $I^-(aq)$ and $H^+(aq)$ ions. Consider this reaction in further detail:

$$H_2O_2 + 2HI \rightarrow 2H_2O + I_2$$

The rate law

$$\text{rate} = k \cdot [H_2O_2]^x \cdot [I^-]^y \cdot [H^+]^z$$

can be modified to

$$\text{rate} = \text{constant} \cdot [H_2O_2]^x$$

if the iodide and hydrogen ion concentrations remain constant. So,

$$\log \text{rate} = \log \text{constant} + x \cdot \log [H_2O_2]$$

If we plot the log (rate) against log $[H_2O_2]$ values at the corresponding times, the gradient will give us the order of reaction, x, with respect to peroxide. (The values of y and z can be determined in a similar manner, by keeping the other concentrations constant.) Table 3.2 gives the data available for the reaction in which the molar concentration of hydrogen peroxide is followed with time.

Table 3.2 Time–concentration data

Time/s	Conc./10^{-2} mol dm^{-3}	log(conc.)	Rate/10^{-4} mol dm^{-3} s^{-1}	log(rate)
30	7.80	−1.11	3.16	−3.50
45	7.35	−1.13	3.02	−3.52
60	6.83	−1.17	2.82	−3.55
75	6.40	−1.19	2.69	−3.57
90	6.00	−1.22	2.51	−3.60
105	5.65	−1.25	2.40	−3.62
120	5.28	−1.28	2.24	−3.65
150	4.61	−1.34	1.95	−3.71
180	4.05	−1.39	1.78	−3.75
210	3.55	−1.45	1.55	−3.81
240	3.10	−1.51	1.38	−3.86
270	2.70	−1.57	1.20	−3.92
300	2.38	−1.62	1.10	−3.96
330	2.05	−1.69	0.93	−4.03
360	1.74	−1.76	0.81	−4.09

The results are plotted in Fig. 3.5. The gradient of the line is determined, tabulated and plotted in Fig. 3.6. The gradient of this line is x, the order of reaction. The reaction is first order with respect to hydrogen peroxide.

Velocity constant

The velocity constant can be determined from a knowledge of the rate, the concentrations of the reactants and the order of reaction. Since

rate $= k \cdot [A]^x \cdot [B]^y$

the velocity constant, k, is given by

$$k = \frac{\text{rate}}{[A]^x \cdot [B]^y}$$

In the peroxide–iodide reaction, the order of reaction for H_2O_2 is 1. If the iodide and hydrogen ion concentrations are constant, then

rate $= k \cdot [H_2O_2] \cdot$ constant

A plot of the rate of reaction against $[H_2O_2]$ will give a straight line of gradient ($k \cdot$ constant). If the value of the constant can be determined (constant $= [I^-]^y \cdot [H^+]^z$), then k can be calculated.

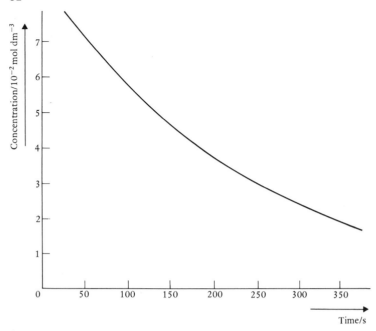

Fig. 3.5 Variation in H_2O_2 concentration with time in its reaction with acid and iodide

The method of calculation is illustrated by the decomposition of an aqueous solution (36 cm^3) of benzene diazonium chloride (10 g dm^{-3}). The reaction is followed by noting the volumes of nitrogen collected after specified time intervals (Table 3.3). The equation of the reaction is

$C_6H_5 \cdot N\!:\!N \cdot Cl \rightarrow C_6H_5Cl + N_2$

Table 3.3 Data for the decomposition of benzene diazonium chloride

Time/m	0	5	10	15	20	25	30	∞
Volume of gas/cm^3	0	15.9	28.0	36.9	43.3	47.6	50.3	58.3
No. of moles remaining/ 10^{-3}	2.56	1.86	1.33	0.94	0.66	0.47	0.35	0.00
Conc/ 10^{-2} mol dm^{-3}	7.12	5.17	3.69	2.61	1.83	1.31	0.97	0.00
Rate/10^{-5} mol dm^{-3} s^{-1}	8.39	5.88	3.79	2.82	2.25	1.44	0.96	–
log (rate)	−4.08	−4.23	−4.42	−4.55	−4.65	−4.84	−5.02	–
log (conc.)	−1.15	−1.29	−1.43	−1.58	−1.74	−1.88	−2.01	–

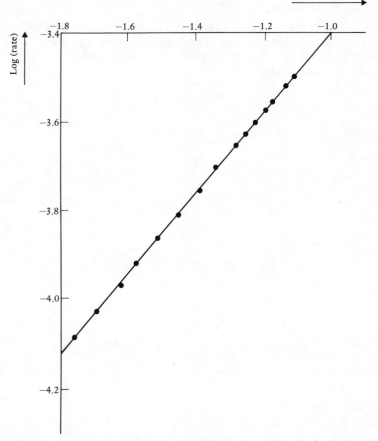

Fig. 3.6 Variation in reaction rate with concentration in the $H_2O_2-H^+-I^-$ reaction

The volume of nitrogen gives the number of moles of $C_6H_5N_2Cl$ decomposed (1 mole of gas occupies 22.4 dm^3), so the concentration remaining may be determined. 36 cm^3 of the original solution contains 2.56×10^{-3} moles (10.0 g dm^{-3} is 7.12×10^{-2} moles per dm^3).

For the reaction,

rate $= k \cdot [C_6H_5N_2Cl]^n$

where n is the order of reaction and assuming that no other reactants are involved. As in the previous example (Figs. 3.5 and 3.6), the rate must be determined from the time–concentration data by drawing tangents to the curve at suitable points. However, it is difficult to draw a tangent very precisely; there is always some uncertainty in its true position. So, the rates obtained by this method are approximate.

64

Converting the rate expression to a logarithmic form gives

log rate = log $k + n \cdot$ log [$C_6H_5N_2Cl$]

Therefore, a graph of the log rate versus log concentration has a gradient of n and the intercept on the log rate axis is log k. The relevant data are plotted in Fig. 3.7. The gradient is 1 and the velocity constant is 1.26×10^{-3} s^{-1} (log $k = -2.90$).

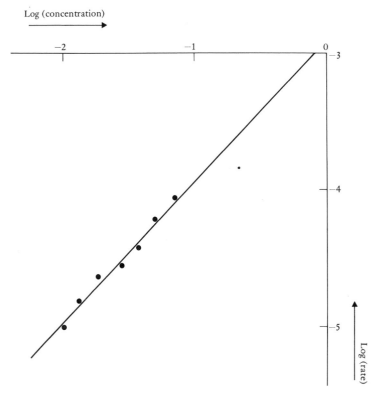

Fig. 3.7   Variation in rate and concentration for the decomposition of benzene diazonium chloride

Multistage reactions

It has been stated (p. 59) that one difficulty of the simple collision theory is that some reactions are more complex than the theory postulates. This can be illustrated by the hydrolysis of 2-chloro-2-methylbutane:

$(CH_3)_3C \cdot Cl + H_2O \rightarrow (CH_3)_3C \cdot OH + H^+(aq) + Cl^-(aq)$

The reaction rate is found to be independent of the concentration of water, and the reaction is first order with respect to the chloroalkane.

Rate = $k \cdot$ [$(CH_3)_3C \cdot Cl$]

It is found that the reaction is a two-stage process:

$$(CH_3)_3 C \cdot Cl \xrightarrow{1} (CH_3)_3 C^+ + Cl^-$$

$$(CH_3)_3 C^+ + H_2O \xrightarrow{2} (CH_3)_3 C \cdot OH + H^+$$

The first reaction is slow with

$$rate_1 = k_1 \cdot [(CH_3)_3 C \cdot Cl]$$

The second stage is fast compared with step 1 and so the overall rate is independent of that step. For example, the rate constant for step 1 is 10^{-2} s^{-1}, whereas that for step 2 is very large compared to this (the reaction is effectively instantaneous), so the overall rate of reaction is determined only by the first step.

The reaction between bromate, bromide and hydrogen ions (experiment 3.9)

$$BrO_3^- + 5 Br^- + 6 H^+ \rightarrow 3 Br_2 + 3 H_2O$$

is even more complex:

$$H^+ + Br^- \underset{}{\overset{1}{\rightleftharpoons}} HBr, \text{ rapid (equilibrium constant, } K_1)$$

$$H^+ + BrO_3^- \underset{}{\overset{2}{\rightleftharpoons}} HBrO_3 , \text{ rapid (equilibrium constant, } K_2)$$

$$HBr + HBrO_3 \underset{}{\overset{3}{\rightleftharpoons}} \text{Complex}$$

$$\rightarrow HOBr + HBrO_2 , \text{ slow (rate constant, } k_3)$$

$$HBrO_2 + HBr \xrightarrow{4} 2HOBr, \text{ fast}$$

$$HOBr + HBr \xrightarrow{5} H_2O + Br_2 , \text{ fast}$$

The steps 1 and 2 have the equilibrium terms

$$K_1 = \frac{[HBr]}{[H^+] [Br^-]} \text{ and } K_2 = \frac{[HBrO_3]}{[H^+] [BrO_3^-]}$$

So, by rearrangement, we have that

$$[HBr] = K_1 \cdot [H^+] [Br^-] \text{ and}$$
$$[HBrO_3] = K_2 \cdot [H^+] [BrO_3^-]$$

For the step 3,

$$rate = k_3 \cdot [HBr] [HBrO_3]$$

on the basis of the simple collision theory. By substitution in this rate equation,

$$
\begin{aligned}
rate &= k_3 \cdot K_1 [H^+] [Br^-] \cdot K_2 \cdot [H^+] [BrO_3^-] \\
&= k_3 \cdot K_1 \cdot K_2 \cdot [H^+]^2 [Br^-] [BrO_3^-] \\
&= k \cdot [H^+]^2 [Br^-] [BrO_3^-]
\end{aligned}
$$

66

When, as in these two examples, one step is slow compared to the rest, this step is known as the **rate determining step.**

Energy profile

During a reaction between two species AB and D, there is an energy change that can be illustrated as in Fig. 3.8. The energy of the products (A + BD)

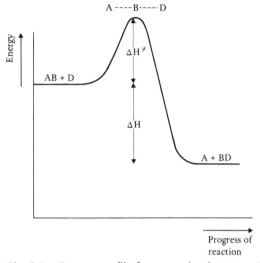

Fig. 3.8   Energy profile for a reaction between AB and D

is less than that of the reactants (AB + D); the reaction is exothermic. Before the products can be formed, however, an energy barrier has to be overcome; this energy is known as the **activation energy.** At this energy maximum the reactants have formed a **transition state,** the **activated complex.**

The transition state is an arrangement of atoms with a higher energy than that of either the reactants or the products. The activation energy is the energy required to reach this state. For the reaction between AB and D, the sequence of events envisaged by this **transition state theory** is:

1.   the approach of AB and D to each other with the correct stereochemistry;
2.   the rearrangement of the electrons leading to a weakening of the bond between A and B and the partial formation of a bond between B and D, giving A ... B ... D;
3.   the rupture of the bond A ... B to give A and BD.

The activation energy arises from the repulsive forces built up between the electrons of the two species; they have to be forced together against this repulsion. The repulsion is overcome by the kinetic energy of the molecules. Figure 3.9 shows diagrammatically the formation of hydrogen iodide from the elements via a transition state.

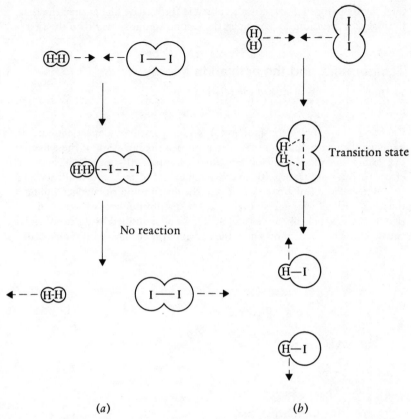

(a)   (b)

Fig. 3.9   Formation of the transition state in the hydrogen–iodine reaction

When a reaction is complex, intermediate compounds are formed. In such cases there are activated complexes in each stage of the reaction. Figure 3.10 illustrates a two-step reaction. The second barrier may be lower

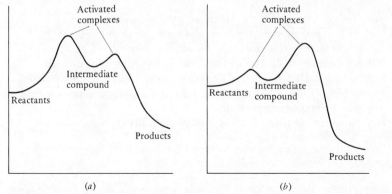

Fig. 3.10   Energy profiles for two-step reactions

68

than the first (Fig. 3.10a); it is also possible for it to be higher than the first. In the latter situation, for example, the second step is slower than the first.

Temperature and the activation energy

In 1889, Arrhenius proposed the relationship

$$k = A \cdot e^{-E/RT}$$

where k is the velocity constant and A is a temperature-independent constant known as the Arrhenius constant. He proposed the inclusion of the exponential term to account for the observation that apparently not all collisions were effective, but only those with an energy in excess of the activation energy, E.

Maxwell and Boltzmann calculated the distribution of energies among the molecules of a gas and showed that the distribution was exponential; that is, the distribution curve (Fig. 3.11) can be described by an exponential equation. Figure 3.11 shows how the distribution changes with temperature.

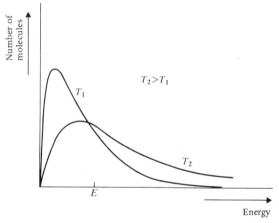

Fig. 3.11   Maxwell–Boltzmann distribution curves

If E indicates the minimum energy required by a molecule to react on collision, then, it is seen, the number of molecules possessing this energy increases as the temperature increases. An activation energy of 25 kJ mol^{-1} means that 1 collision in 22 200 will be of sufficient energy to react (the exponential factor is 4.5×10^{-5}); a 10 per cent drop to 22.5 kJ mol^{-1} raises the ratio to 1 in 8 100 (exponential term 1.2×10^{-4}). So, the rate of reaction increases with increasing temperature.

The Arrhenius equation shows the relationship between the velocity constant and temperature. It may be arranged to give

$$\log_e k = \log_e A - E/_{RT}$$

In the experiment described above (benzene diazonium chloride decomposition, p. 62), the velocity constant varies with temperature as shown in

Table 3.4. A plot of $\log_{10} k$ against $1/_T$ gives a line with gradient $-E/2.303R$ (Fig. 3.12). The value of the activation energy is 111 kJ mol^{-1}.

Table 3.4 Variation in the rate constant with temperature

Temp/K	k/s^{-1}	$\log k$	$10^3/T$
288	9.30×10^{-6}	−5.03	3.47
293	2.01×10^{-5}	−4.70	3.41
298	4.35×10^{-5}	−4.36	3.36
303	9.92×10^{-5}	−4.00	3.30
308	2.07×10^{-4}	−3.68	3.25
313	4.28×10^{-4}	−3.37	3.19
318	8.18×10^{-4}	−3.09	3.14

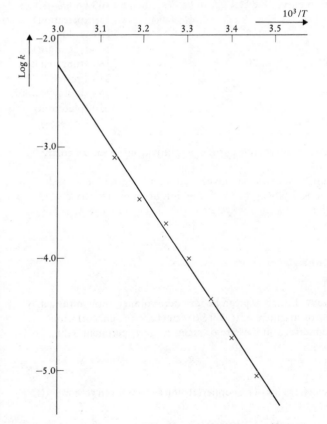

Fig. 3.12 Variation in the rate constant with temperature

It will be found, by comparison of the data in Table 3.4, that an increase of 10° in this reaction increases the reaction rate by approximately

4.7. This clearly is an approximation but it is a useful guide. A doubling of the rate by a rise of $10°$ corresponds to an activation energy of about 55 kJ mol^{-1} (at 300 K).

Catalysis and the activation energy

It is found that catalysts lower the value of the activation energy as determined experimentally. Table 3.5 gives examples of some catalytic effects.

Table 3.5   Activation energy for some catalysed decompositions

Reaction	Activation energy/kJ mol^{-1}	
	Uncatalysed	Catalysed
$2HI \rightarrow H_2 + I_2$	184	105 (gold)
		59 (platinum)
$2N_2O \rightarrow 2N_2 + O_2$	245	121 (gold)
		136 (platinum)
$2NH_3 \rightarrow N_2 + 3H_2$	330	162 (tungsten)
		213 (iron)
$2(C_2H_5)_2O \rightarrow C_2H_6 +$		144 (iodine vapour)
$2CO + 4CH_4$	224	50 (platinum)
$2H_2O_2 (aq) \rightarrow 2H_2O + O_2$	75	21 (catalase)

The catalyst provides an alternative path, a path involving an intermediate compound (Fig. 3.10).

The mechanism of catalysis may involve either adsorption on a solid surface (heterogeneous catalysis) or the formation of an intermediate compound (usually a homogeneous system). The latter route can be represented by the steps

$A + C \rightarrow AC$
        (intermediate)
$AC + B \rightarrow$ Products + C

where C is the catalyst. Either step can be rate-determining, though they may be of a similar order of magnitude. It will be seen that the concentration of the catalyst has an effect on the rate of reaction. In experiment 3.11, the reaction

$$2Fe^{3+}(aq) + 2I^-(aq) \rightarrow 2Fe^{2+}(aq) + I_2(aq)$$

is catalysed by copper(II) ions. The copper(II) ions are reduced to copper(I) ions by iodide ions:

$$2Cu^{2+}(aq) + 2I^-(aq) \rightarrow 2Cu^+(aq) + I_2(aq)$$

The copper(I) ions are oxidised to copper(II) ions by iron(III) ions, which, in turn, are reduced to Fe^{2+}. More copper(I) ions are generated by the iodide ions.

Many industrial processes involve homogeneous catalysis. For example, the carbonylation of an alkene produces an aldehyde in the presence of a cobalt compound as catalyst.

$$CH_2 : CH_2 + CO + H_2 \xrightarrow{\text{Co compound}} CH_3 CH_2 CHO$$

Surface catalysis involves the adsorption of the reactant molecules, reaction between the reactants and then desorption of the products. The exact mechanisms vary according to the system considered. Adsorption may be merely physical (that is due to weak interactions such as the van der Waals' forces) or chemical ('chemisorption'). For example, when nitrogen is chemisorbed on a metal, the electron density between the nitrogen atoms is reduced due to the formation of bonds between the nitrogen atoms and the metal. As a result, the nitrogen atoms are more reactive than prior to adsorption.

It is found that metals can be roughly classified according to their ability to chemisorb gases (Table 3.6). The effectiveness of a catalyst is determined

Table 3.6 Catalytic activity of metals

Metals	Gases						
	O_2	C_2H_2	C_2H_4	CO	H_2	CO_2	N_2
Ti, V, Cr, Fe	√	√	√	√	√	√	√
Co, Ni	√	√	√	√	√	√	×
'Pt metals'	√	√	√	√	√	×	×
Mn, Cu, Au, Al	√	√	√	√	×	×	×
Zn, Ag, Mg,							
p block metals	√	×	×	×	×	×	×

√ strong, × weak

by the balance of several factors:

(a) adsorption must not be too strong, or else one strong bond is merely replaced by another and the activation energy is not reduced;

(b) adsorption must not be too weak, or else the electron density is not disturbed and the gas molecules are ineffectively held by the metal;

(c) desorption must be ready – a strongly held product will prevent further reaction.

While the rate of reaction is increased by an increase in temperature, there is an optimum temperature for a heterogeneous catalytic reaction. If the temperature is too high, the molecules have too much kinetic energy and collide with the catalyst with too much energy to be adsorbed. A typical curve is that shown in Fig. 3.13. The maximum in the curve represents the balance between the normal rate–temperature relationship and the decreasing tendency to adsorption as the temperature increases. Past the maximum point the latter effect is predominant and the overall reaction rate decreases.

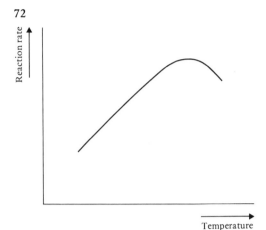

Fig. 3.13 Variation in reaction rate with temperature for a catalytic reaction

Enzymes

Biological reactions are catalysed by enzymes. Enzymes are catalysts produced by the living cell. Their activity is greater than that of simple catalysts. They are very sensitive to temperature and pH. For example, the optimum temperature is between 300 and 310 K; above that temperature they are destroyed. Ptyalin, in the saliva, acts on the substrate to produce maltose; the optimum pH is 6.7. Another digestive enzyme, pepsin, in the stomach acts at pH 1.4 on the proteins to produce peptides. Further in the digestive system, the duodenum, the pH is high and trypsin acts on the peptides to produce amino acids.

In contrast to heterogeneous catalysts in which the whole of the surface of the particle is apparently available for catalytic activity, the activity of an enzyme is concentrated in a few localised positions. These places are known as the **active sites.** The enzymes are also highly specific – each enzyme acts on only one substrate. This appears to be due to stereochemical requirements. The mechanism involves a 'lock and key' effect (Fig. 3.14).

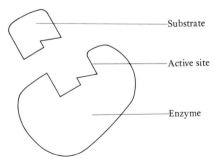

Fig. 3.14 Enzyme 'lock and key' mechanism

The only substrate affected by the enzyme is the one that fits into the active site like a key into a lock. The reaction is even specific enough to differentiate between isomers.

Because of their highly specific character, enzymes are now being introduced into chemical syntheses and are being used in analytical techniques.

Industrial applications

Any chemical process which is required to be economic must take into account not only the yield, but also the rate at which that yield is obtained. The survival of an industry is based on the amount produced at the end of the day — a 50 per cent conversion at the end of the day is much better than 100 per cent at the end of the century! We can apply these considerations to two industrial processes.

(a) Manufacture of sulphuric acid

The essential step in the process is the reaction of sulphur dioxide with oxygen:

$$2SO_2 + O_2 \rightleftharpoons 2SO_3, \Delta H = -98.5 \text{ kJ mol}^{-1}$$

Figure 3.15 shows how the yield of sulphur trioxide varies with temperature, based on an 8 per cent initial concentration of SO_2 which results in a

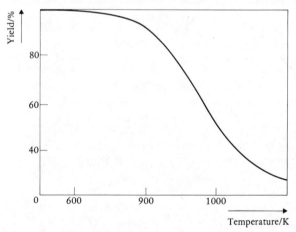

Fig. 3.15 Relationship of SO_3 yield to temperature

high conversion rate for this gas. It is found that in excess of 90 per cent can be obtained at temperatures below 800 K.

The reaction can be catalysed by platinum or vanadium(V) oxide. Table 3.7 lists some of the typical activation energies for these catalysed processes. While platinum is the most effective, having the lowest activation energy, it is also the most easily poisoned. The vanadium(V) oxide is, there-

Table 3.7 Activation energies of SO_3 synthesis

Catalyst	Activation energy/kJ mol^{-1}
Pt	42
Pt on SiO_2	96
V_2O_5, K_2SO_4, SiO_2	113
Commercial V_2O_5	125
Pure V_2O_5	151

fore, a more realistic choice. The alkaline V_2O_5 catalytic system (activation energy, $+113$ kJ mol^{-1}) has the lower activation energy barrier of the types specified and is the one used industrially.

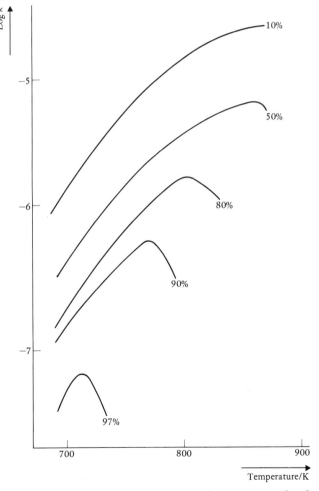

Fig. 3.16 Variation in rate constant with temperature for the SO_3 synthesis using V_2O_5 catalyst

Figure 3.16 shows how the rate constant varies with temperature over solid V_2O_5 using the 8 per cent SO_2 input. The fractions shown indicate the fraction of conversion achieved. Since the rate is proportional to the rate constant, it is observed that there is a maximum rate in each case (compare Fig. 3.13) and that this maximum lies in the range 700–1000 K. While a 97 per cent conversion occurs at 700 K, it is ten times slower than the 90 per cent yield at 775 K. At the former temperature, the relative rates of reaction for the varying particle diameters are:

0.67 mm 4.3
1.38 mm 2.9
8.00 mm 1.0

The smaller particles produce higher rates (compare p. 56).

The modern industrial process uses a homogeneous catalyst system, $V_2O_5 - K_2SO_4 - SiO_2$ (see Table 3.7). This catalyst is exposed to the reaction

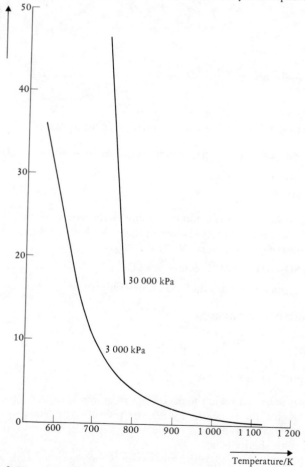

Fig. 3.17   Variation in yield of ammonia with temperature

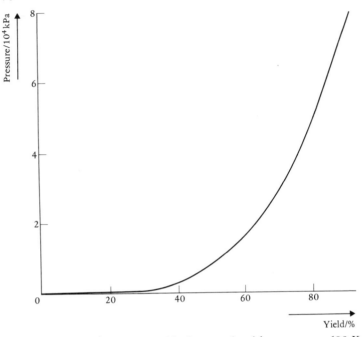

Fig. 3.18 Variation in the yield of ammonia with pressure at 600 K

mixture and the sulphate is converted to a pyrosulphate by some of the SO_3 formed initially:

$$SO_4^{2-} + SO_3 \rightarrow S_2O_7^{2-}$$

The pyrosulphate dissolves the vanadium(V) oxide on the surface of the silica support and the reactant gases dissolve in the melt. Some of the vanadium(V) is reduced to vanadium(IV) by the SO_2:

$$2V^V \text{ complex} + SO_2 + O^{2-} \rightarrow 2V^{IV} \text{ complex} + SO_3$$

Reaction with oxygen regenerates the vanadium(V) species.

(b) Manufacture of ammonia

$$\tfrac{1}{2}N_2 + \tfrac{3}{2}H_2 \rightleftharpoons NH_3$$

$$\Delta H = -46.2 \text{ kJ mol}^{-1} \text{ at 298 K}$$

Figure 3.17 shows the variation in the percentage of ammonia with temperature and Fig. 3.18 gives the variation in yield with pressure, at a temperature of 600 K. From these graphs it is apparent that low temperatures and high pressures give the best yields.

The best catalysts appear to be those based on the iron group metals with a d^6 electronic structure (Table 3.8). The relationship of catalytic activity to

Table 3.8   Activation energies for catalysed NH_3 synthesis

Catalyst	Activation energy/kJ mol⁻¹
None	283.8
Osmium	134.8
Tungsten	115.6
Molybdenum	176.2
Iron	167.0

the enthalpy of adsorption of nitrogen on d-block metals is shown in Figs. 3.19a and 3.19b. The maximum rate occurs where the energy of

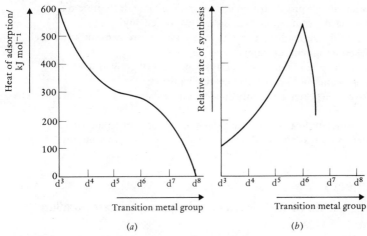

(a)   (b)

Fig. 3.19   Catalytic effect and d-electron configuration

adsorption is neither too high nor too low (see p. 71). The commercial catalyst is based on iron oxide, the activity of which is enhanced by the use of alkaline oxides. The catalyst is poisoned by sulphur, phosphorus, carbon monoxide, water and carbon dioxide and so these must be removed from the feedstock.

The process operated in this country, the Haber process, uses pressures of up to 35.5×10^3 kPa and temperatures of approximately 600 K. Yields of between 15 and 20 per cent are obtained. The Claude process, operated in other parts of the world, uses pressures of up to 10^5 kPa to improve the yield.

Summary

At the conclusion of this chapter, you should be able to:

1. state that the rate of reaction is the rate at which a product is formed or a reactant consumed;

2. determine the rate of reaction from a concentration–time graph;
3. state that the rate of reaction decreases as the reactants are consumed;
4. describe the factors affecting reaction rate;
5. define the term catalysis;
6. explain the term autocatalysis;
7. describe some of the techniques used to follow the progress of a reaction;
8. outline the principles of collision theory and its limitations;
9. state the law of mass action;
10. define the order of reaction and velocity constant;
11. calculate the order of a chemical reaction from suitable experimental data;
12. explain the concept of a rate determining step;
13. draw an energy profile for an exothermic or an endothermic reaction, indicating the transition state and activation energy;
14. calculate the activation energy from a table of rate constants and their corresponding temperatures;
15. explain the phenomenon of catalysis in terms of activation energy, homogeneous and heterogeneous mechanisms;
16. state that enzymes are highly specific and sensitive catalysts produced by a living cell;
17. discuss the applications of the principles of kinetics to industrial processes such as the production of sulphuric acid and ammonia.

Experiments

3.1 Investigation of the kinetics of the permanganate–oxalate reaction

This reaction can be followed conveniently in a colorimeter as the intensely coloured permanganate ions (MnO_4^-) are converted to colourless manganese(II) ions.

$$2MnO_4^-(aq) + 5C_2O_4^{2-}(aq) + 16H^+(aq) \rightarrow$$

$$2Mn^{2+}(aq) + 10CO_2(g) + 8H_2O(l)$$

The colorimeter is fitted with a filter which is selected to give the maximum value for the absorbance (see p. 140). Set the colorimeter to give the minimum absorbance with distilled water. It is standardised by determining the absorbance for a series of standard permanganate solutions (e.g. $0.25, 0.50, 1.00, 2.00 \times 10^{-4}$ mol dm^{-3}). Plot a graph of the absorbance against concentration. Do not adjust the instrument for the rest of the experiment.

Prepare a mixture of 0.10 M oxalate (ethan-1, 2-dioate) solution (10.0 cm^3) in 1.2 M sulphuric acid. Put 0.02 M permanganate solution (0.2 cm^3) into the cell of the colorimeter, add the oxalate solution, mix thoroughly and start a stop clock. Take readings from the colorimeter at 10 s intervals until the reaction is complete.

Convert the absorbance readings to concentrations of permanganate by means of the calibration curve. Plot the concentration of permanganate against reaction time. Discuss the significance of the shape of the graph. By determining the gradient of the graph at appropriate points, calculate the rate of reaction at these times.

3.2 Investigation of the factors affecting the rate of the peroxodisulphate–iodide reaction

$$S_2O_8{}^{2-} + 2I^- \rightarrow 2SO_4{}^{2-} + I_2$$

The reaction is performed using varying concentrations, varying temperatures and a catalyst. In order to ensure that the reaction is followed to exactly the same point in each case (so making it possible to compare the reaction rates), a standard amount of sodium thiosulphate solution is added. When the thiosulphate has been consumed,

$$2S_2O_3{}^{2-} + I_2 \rightarrow 2I^- + S_4O_6{}^{2-}$$

the iodine is liberated; its presence is shown as a blue colour in the presence of starch.

In a boiling tube place 0.5 M KI(aq) (20 cm^3) and 0.01 M Na$_2$S$_2$O$_3$(aq) (10 cm^3). Another tube should contain 0.01 M K$_2$S$_2$O$_8$(aq) (20 cm^3) and a few drops of starch solution. Bring both solutions to 20 °C, mix the contents and start a stop clock. Note the time taken for the appearance of the blue colour.

Repeat the reaction using chilled solutions (0–5 °C), and after warming to 30 °C. Other temperatures may be used if time permits.

The first run should be repeated using 10 cm^3 of the peroxodisulphate and 10 cm^3 water with the starch. This gives half the concentration of this reactant.

Repeat the first run, but include 1–2 drops of an aqueous solution of iron(II) sulphate in the first tube.

Account for the results. How is the rate of reaction affected by the concentration of the peroxodisulphate and by the temperature?

3.3 Measurement of reaction rates by gas evolution

The apparatus consists of a conical flask, containing the reactants, a gas delivery tube from the flask to a beaker of water and an inverted burette to collect the gas. The burette should be filled with water.

In the flask place 30 per cent hydrochloric acid (100 cm^3) and zinc foil (0.5 g); connect the apparatus and start the timing. Record the time taken to displace the water to the 50, 40, 30, 20, 10 and 0 cm^3 marks on the burette.

Repeat the experiment using zinc wool and zinc dust. The first run should be repeated using 15 per cent acid, and then at 10 °C higher.

Plot the results (volume against time) and comment on the graphs.

3.4 Further investigation of the permanganate–oxalate reaction

Prepare a mixture of 0.2 M oxalic acid (100 cm^3), 2 M sulphuric acid (5 cm^3) and water (95 cm^3). Add 0.02 M KMnO$_4$(aq) (50 cm^3) and start a stop clock. Shake the mixture for 30 seconds. After 1 minute withdraw an aliquot

(10 cm^3) and pour it into a small flask containing $0.1 M$ KI (10 cm^3). The I^-(aq) reacts instantaneously with the MnO_4^-, so stopping the $MnO_4^- - C_2O_4^{2-}$ reaction in that portion, and releases iodine, the quantity of which is proportional to the unreacted permanganate.

Remove further aliquots at 3 minute intervals for a period of up to 30 minutes. Titrate the liberated iodine in each case against $0.1 M$ thiosulphate. Plot your titre values against time.

Repeat the above procedure, but use a mixture of $0.2 M$ oxalic acid (100 cm^3), $0.2 M$ manganese(II) sulphate (15 cm^3), $2 M H_2SO_4$(aq) (5 cm^3) and water (80 cm^3). The samples should be withdrawn at 3 minute intervals up to 15 minutes. Comment on the contrasting graphs of these two runs.

3.5 Use of conductance to follow the rate of saponification of an ester

$$CH_3COOCH_2CH_3(aq) + OH^-(aq) \rightarrow CH_3CH_2OH(aq) + CH_3COO^-(aq)$$

The reaction may be followed by conductance measurements because the molar conductances of the hydroxide and ethanoate ions are different (p. 133).

Place $0.05 M$ sodium hydroxide solution (50 cm^3) in a beaker and determine its conductance. Add $0.05 M$ ethyl ethanoate solution (50 cm^3), start the stop clock and stir the mixture. The mixture should be stirred throughout the experiment. Initially, record the conductance every minute (up to 10 minutes) and then at 2 minute intervals (up to 30 minutes). Plot the conductance readings against time.

3.6 Measurement of reaction rates by precipitation methods

$$S_2O_3^{2-}(aq) + 2H^+(aq) \rightarrow H_2O(l) + SO_2(aq) + S(s)$$

Prepare $1.5, 1.0, 0.5, 0.1, 0.05 M$ sodium thiosulphate solutions.

To portions of thiosulphate (5 cm^3), add $0.5 M$ hydrochloric acid (5 cm^3) and start a stop clock. Note the time taken for the first appearance of a sulphur precipitate. (Any intensity of precipitate may be taken as long as the same point is used in each case.)

Plot $1/t$ against the concentration of thiosulphate solution. (The reaction rate is inversely proportional to the time.)

3.7 Determination of the order of reaction in the Harcourt—Esson reaction

$$H_2O_2(aq) + 2H^+(aq) + 2I^- \rightarrow I_2(aq) + 2H_2O(l)$$

The progress of the reaction is followed by determining the time taken to liberate equal amounts of iodine. This is monitored by the reaction of iodine with thiosulphate ions:

$$2S_2O_3^{2-}(aq) + I_2(aq) \rightarrow 2I^-(aq) + S_4O_6^{2-}(aq)$$

Set up a burette with $0.1 M$ thiosulphate solution. Prepare a mixture of 50 per cent sulphuric acid (10 cm^3), starch (10 cm^3) and '1-volume' hydrogen peroxide solution (10 cm^3) and put it in a conical flask. In another flask, place 0.06 M potassium iodide solution (200 cm^3). Both flasks should be placed in a constant temperature bath. Add 1 cm^3 of the thiosulphate solution to the iodide and pour in the peroxide mixture. Start a stop clock. Note the time when the blue colour appears and immediately stir in a further 1 cm^3 portion of thiosulphate. Repeat the procedure up to a total of 14 cm^3 thiosulphate.

These volumes of thiosulphate are a measure of the amount of H$_2$O$_2$ consumed in the reaction. The amount of H$_2$O$_2$ remaining (which determines the reaction rate) is proportional to the difference $(V_t - V_\infty)$, where V_t is the volume of thiosulphate at time t, and V_∞ is the volume of thiosulphate required when all the peroxide has been consumed.

The volume V_∞ can be determined as follows. To a sample of the peroxide solution (10 cm^3), *carefully* add concentrated sulphuric acid (10 cm^3) followed by potassium iodide (8 g). The solution may be diluted to dissolve any excess solid, and it is titrated against the thiosulphate solution in the presence of starch indicator.

Plot V_t against t; draw tangents at suitable points to determine the rate of reaction. Plot log rate v. log $(V_t - V_\infty)$ and the gradient will give the order of reaction with respect to hydrogen peroxide:

$R = k$ [H$_2$O$_2$]^x [I$^-$]^y [H$^+$]^z

$R = k'$ [H$_2$O$_2$]^x since [I$^-$] and [H$^+$] are constant,

so, $\log R = \log k' + x \cdot \log$ [H$_2$O$_2$]

3.8 Determination of the order of reaction in the permanganate–oxalate reaction

Repeat experiment 3.4, using 0.05, 0.025, 0.01, 0.005 M MnO$_4^-$(aq).

Determine the gradient of each curve at $t = 0$ and plot the log (gradient) against log [MnO$_4^-$]. The gradient of the latter graph gives the order of reaction with respect to permanganate.

3.9 Examination of the bromate–bromide reaction

BrO$_3^-$(aq) + 5Br$^-$(aq) + 6H$^+$(aq) → 3Br$_2$(aq) + 3H$_2$O

The rate of reaction is determined by the time taken to produce a fixed amount of bromine. This is specified by the addition of a specific quantity of phenol (hydroxybenzene):

When the phenol has been consumed, the bromine decolorises methyl red indicator. So, the reaction is run in each case until the red colour is discharged. All the solutions used should be brought to some suitable temperature (e.g. 20 °C) before the reaction is commenced.

In the first run, prepare a mixture of 0.005 M $KBrO_3$ (100 cm^3), 0.000 05 M phenol (10 cm^3), 2 M H_2SO_4 (30 cm^3) and water (110 cm^3). In a series of beakers place 10.0, 8.0, 6.0, 4.0, and 2.0 cm^3 0.01 M KBr(aq) respectively, and add sufficient water in each case to bring the total volume to 10 cm^3. In each case, add the bromide solution to 25 cm^3 of the bromate solution containing 2–3 drops of methyl red and start a stop clock. Note the time at which the solution becomes colourless in each case. Plot the concentration of the bromide against the time. The gradient of the graph at any point gives the rate of the reaction for the corresponding concentration. Determine the rate at a series of concentrations and plot the log (rate) v. log (concentration). The relationship between the rate and the concentration is given by

$$\text{rate} = \text{constant } [Br^-]^x$$

or, $\log (\text{rate}) = \log (\text{constant}) + x \cdot \log [Br^-]$

The gradient of this new plot gives x, the order of reaction for Br^- ions.
The second run should use the following solutions:

0.005 M $KBrO_3$ /cm^3:	10.0	8.0	6.0	4.0	2.0
Water/cm^3:		11.0	13.0	15.0	17.0	19.0

To each add 0.000 05 M phenol (1 cm^3), 2 M H_2SO_4 (3 cm^3) and 2–3 drops of methyl red. The reaction is initiated by the addition of 0.01 M KBr (10 cm^3) to each. Again, the reaction is timed to the complete discharge of the red colour. Plot $[BrO_3^-]$ against time and determine the rate at various concentrations. Plot log (rate) against log $[BrO_3^-]$ and determine the order of reaction with respect to bromate ions, as done for Br^- ions.

The third run uses varying quantities of acid. Prepare the following solutions:

0.01 M H_2SO_4 /cm^3:	10.0	8.0	6.0	4.0	2.0
Water/cm^3:	0.0	2.0	4.0	6.0	8.0

To each add 0.000 05 M phenol (1 cm^3), 0.2 M $KBrO_3$ (10 cm^3) and 2–3 drops of methyl red. The reaction is timed from the addition of 0.4 M KBr (15 cm^3). Plot $[H^+]$ against time and so log (rate) against log $[H^+]$. Determine the order of reaction with respect to hydrogen ions, as for the previous two ions.

3.10 Determination of the activation energy of the peroxodisulphate–iodide reaction

Use the results of experiment 3.2 for the run at several temperatures (e.g. 0, 20, 30 and 40 °C).

$$k = A e^{-E/RT}$$

So, $\log_e k = \log_e A - \dfrac{E}{RT}$

or $\log_{10} k = \log_{10} A - \dfrac{E}{2.3R} \cdot \dfrac{1}{T}$

Since the $[S_2O_8^{2-}]$ and $[I^-]$ are identical in each case,

Rate $\propto k$ at each temperature.

As the reaction runs to the same point (concentration of I_2 is identical),

Rate $\propto 1/t$

So, a plot of $\log_{10}(1/t)$ v. $1/T$ gives a straight line of slope $-E/2.3R$, where $R = 8.314$ J deg^{-1} mol^{-1}.

3.11 Investigation of the catalytic action of copper(II) ions

(a) In a flask place 0.015 M $K_2Cr_2O_7$ (10 cm^3), glacial ethanoic acid (2.5 cm^3), water (10 cm^3) and starch solution (1 cm^3). Add 5 per cent KI (aq) (30 cm^3), start a stop clock and immediately add 0.10 M thiosulphate solution (5 cm^3). Note the time taken for the blue starch–iodine colour to form (i.e. after the consumption of the thiosulphate).

$$Cr_2O_7^{2-} + 14H^+ + 6I^- \rightarrow 2Cr^{3+} + 7H_2O + 3I_2$$

$$2S_2O_3^{2-} + I_2 \rightarrow 2I^- + S_4O_6^{2-}$$

Repeat the reaction, but add 0.001 M $CuSO_4$(aq) (1 cm^3) to the dichromate mixture.

(b) Add 0.10 M sodium thiosulphate solution (5 cm^3) to 0.10 M iron(III) chloride solution (5 cm^3) and note the time taken for the initial colour to be discharged. Repeat in the presence of Cu^{2+}(aq) as in (a). The reaction can be repeated using CNS^-(aq) instead of Cu^{2+}(aq). Can you account for the action of the thiocyanate?

$$Fe^{3+} + 2S_2O_3^{2-} \xrightarrow{\text{fast}} [Fe(S_2O_3)_2]^- \xrightarrow{\text{slow}} Fe^{2+} + S_4O_6^{2-}$$

Copper(II) ions are reduced, by iodide or thiosulphate, to copper(I) which then reduces the metal ions.

3.12 Investigation of an enzyme catalysed reaction

The apparatus required is as for experiment 3.3. Prepare a series of hydrogen peroxide solutions varying from 100-volume to 10-volume concentration. The catalyst consists of yeast (0.05 g) suspended in water (5 cm^3). A series of buffer solutions ranging from pH 2.4–8.0 are required (citric acid–phosphate). All solutions should be brought to a common temperature (e.g. 20 °C).

Using the 20-volume peroxide (10 cm^3) and each of the buffer solutions in turn (5 cm^3), add the yeast suspension (5 cm^3) and note the volume of gas at every 30 seconds. Plot time against pH and determine the optimum pH.

Using the optimum pH, repeat the run using varying concentrations of peroxide. Plot the time against concentration.

The optimum temperature can be found using 20-volume peroxide (10 cm^3), optimum pH (5 cm^3 buffer), and yeast suspension (5 cm^3). Plot time against temperature.

Repeat the run at the optimum temperature but using varying amounts of yeast (0.01–0.10 g in 5 cm^3 water). Plot the time v. enzyme concentration. Comment on the result.

84

References

An Introduction to Reaction Kinetics; M. A. Atherton and J. K. Lawrence (Longman, 1970).
HABER (a computer tutorial exercise) (Edward Arnold, 1977).
Principle of Catalysis; G. C. Bond (RIC Monograph No. 7, 1963).

Films

Introduction to reaction kinetics (CHEM Study).

Questions

1. Radioactive iodine (^{128}I) decays by β-radiation. The concentration of the isotope can be determined by measurement of its activity. The data indicate the activity of ^{128}I after various time intervals.

Activity/counts s^{-1}:	142.5	117.5	77.0	47.0	28.7	17.0	10.8
Time/s:	500	1 000	2 000	3 000	4 000	5 000	6 000

   Plot the activity against time. Determine the gradient (= rate of decay) at 1 000 s, 3 000 s and 5 000 s. Plot the rate against the time.

2. It is reported that silver ions catalyse the oxidation reactions of the peroxodisulphate ion. Using a suitable example, describe in detail how you would test this claim.

3. Potassium forms a blue solution with liquid ammonia. The metal reacts slowly with the solvent and is accompanied by a loss of colour. The reaction may be followed colorimetrically. The data show how the concentration changes with time. Determine the order of reaction with respect to potassium.

Time/minutes	0	150	275	400	600	750	1 300
Concentration/10^{-4} mol dm^{-3}		11.3	9.80	8.77	7.85	6.45	5.59	3.36

4. Nitrogen(V) oxide decomposes at 320 K according to the equation

$$N_2O_5 \rightarrow 2NO_2 + \tfrac{1}{2}O_2$$

   The following data were obtained in a typical experimental run; show that the reaction is first order with respect to nitrogen(V) oxide.

Time/minutes	0	10	20	40	60	80	100	120
Pressure of N_2O_5/kPa	46.4	32.9	24.6	14.0	7.7	4.4	2.4	1.3

5. The rate constant for a reaction was determined at several temperatures:

Temperature/K	285	290	298	306
Rate constant/dm^3 mol^{-1} s^{-1}	0.010 7	0.028 2	0.126	0.525

   Calculate the energy of activation for the reaction.

Part 2

Laboratory techniques

Chapter 4

Separation techniques

Background reading

The background to this chapter is covered in *Fundamentals of Chemistry*, Chapter 6.

We are inevitably concerned with methods of separating chemical substances in order to obtain one substance in a pure state, whether for a preparation or for analytical purposes. This chapter reviews some of the separation techniques commonly used.

Ion exchange resins

An ion exchange resin is a macromolecular substance with ionic units which can be exchanged with ions in a surrounding solution. Many common resins are prepared from polystyrene. Styrene, $C_6H_5 \cdot CH=CH_2$, is polymerised in the presence of a catalyst (e.g. benzoyl peroxide):

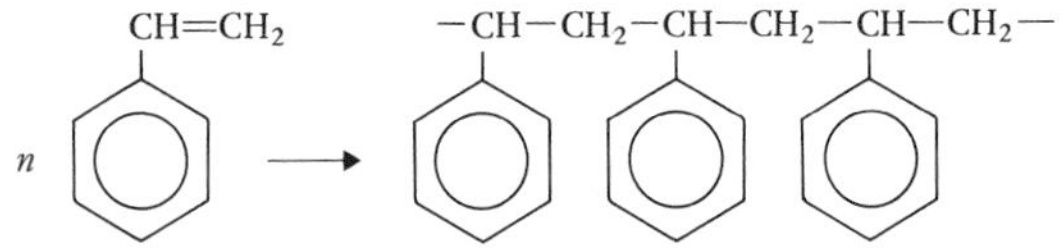

The characteristics of this polymer (thermoplastic, soluble in organic solvents) are altered by cross-linking the chains. This is achieved by introducing a small amount of 1,4-diethenylbenzene into the reaction mixture:

$$CH=CH_2 \quad CH=CH_2 \quad -CH-CH_2-CH-CH_2-CH-CH_2-$$

(polymerization scheme of styrene and divinylbenzene)

The polymer is formed as small beads. It can be sulphonated with concentrated sulphuric acid (Fig. 4.1). Substitution may be in either the 2- or 4-ring positions.

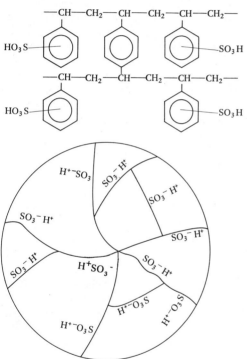

Fig. 4.1 A strongly acidic resin structure

A basic group can be introduced by the Friedel–Craft's reaction of polystyrene with chloromethoxymethane, followed by reaction with an amine, such as trimethylamine (N, N-dimethylaminomethane).

$$-CH-CH_2-CH-CH_2- \qquad -CH-CH_2-CH-CH_2-$$

$$\xrightarrow[CH_3OCH_2Cl]{Al_2Cl_6} \qquad \xrightarrow{N(CH_3)_3}$$

$$CH_2Cl \qquad CH_2Cl$$

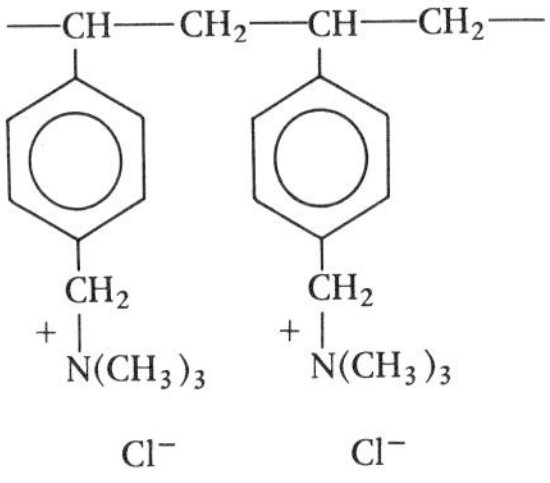

$$-CH-CH_2-CH-CH_2-$$

The resins described are strongly acidic and strongly basic respectively. Carboxylic acid groups provide weak acid resins. The pH of these can be buffered using the appropriate concentration of the salt solution (see p. 35). Primary and secondary amine groups generate weak base exchangers.

There are also some naturally-occurring ion exchange materials. These are aluminosilicates. Silicates are based on polymeric SiO_4^{4-} units; aluminosilicates are formed by the replacement of some silicon atoms with aluminium atoms. A variety of such compounds are naturally occurring – clays, zeolites and felspathoids.

Hydrous oxides, formed by the action of aqueous ammonia on metal salt solutions, are filtered and dried at low temperatures to generate further resins. The three-dimensional networks are formed by hydroxide groupings interacting with the loss of water:

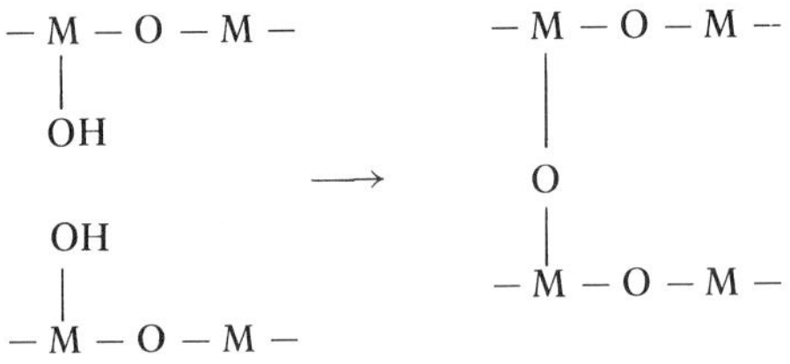

Residual hydroxide groups may be either basic (ionising to M^+ and OH^-) or acidic (MO^- and H^+), depending on the pH of the medium.

An equilibrium is set up between ions on the resin and those in solution. The ions in the mobile medium are known as counter ions. Where A and B are monovalent ions and $\textcircled{R}$ is the resinous material,

$$\textcircled{R}-A + B(aq) \rightleftharpoons \textcircled{R}-B + A(aq)$$

The equilibrium constant or selectivity coefficient, $K_{A/B}$, is given by

$$K_{A/B} = \frac{[A(aq)] [\,\textcircled{R}-B]}{[\,\textcircled{R}-A] [B(aq)]}$$

The selectivity of a resin for an ion is given by the distribution coefficient K_d. For example, for substance A

$$K_d(A) = \frac{\text{Concentration of A in the resin}}{\text{Concentration of A in the solution}}$$

So, the equilibrium constant $K_{A/B}$ is related to the distribution coefficients as

$$K_{A/B} = \frac{K_d(B)}{K_d(A)} = \frac{[\text{\textcircled{R}}-B]}{[B(aq)]} \cdot \frac{[A(aq)]}{[\text{\textcircled{R}}-A]}$$

Table 4.1 gives some typical values of the distribution coefficients.

Table 4.1 Distribution coefficients for the sodium group ions from 0.1 M NH_4NO_3 solution

Cation	$K_d/cm^3 g^{-1}$		Selectivity coefficient on Dowex 50
	Ammonium phosphomolybdate	Dowex 50	
Li^+		14	1.00
Na^+	0	26	1.88
K^+	3.4	46	2.63
Rb^+	230	52	2.89
Cs^+	6 000	62	2.91

The *rate* of exchange is dependent on the nature of the resin and on the size and concentration of the ions being exchanged. The resin takes up water on its surface due to the hydrophilic (water-attracting) nature of the substituent groups.

The rate determining steps in the ion exchange process are: (*a*) diffusion of the ions across the adsorbed liquid film; and (*b*) diffusion of the ions within the resinous material. The relative importance of these steps can be controlled. The film diffusion process becomes the controlling factor when small ions, dilute solutions, small exchanger particles and resins with a low degree of cross-linking are used. The converse conditions favour particle diffusion as the rate determining step.

In dilute solutions the exchange potential of a resin (the ease with which an ion in solution displaces the absorbed ion on the resin) increases with increasing valency:

$$Fe^{3+} > Al^{3+} > Ba^{2+} > Ca^{2+} > Ni^{2+} = Cu^{2+} > Mg^{2+} > Ag^+ > NH_4^+ = K^+ > Na^+ > H^+$$

Concentrated solutions show the reverse effect. Typical separation reactions are

$$R \cdot SO_3^- H^+ + Na^+(aq) \rightleftharpoons R \cdot SO_3^- Na^+ + H^+(aq)$$

$$R \cdot SO_3^- Na^+ + K^+(aq) \rightleftharpoons R \cdot SO_3^- K^+ + Na^+(aq)$$

$$R \cdot NH_3^+ Cl^- + FeCl_4^-(aq) \rightleftharpoons R \cdot NH_3^+ FeCl_4^- + Cl^-(aq)$$

A glass tube, about 30 cm long and fitted with a tap (Fig. 4.2), is filled with an ion exchange resin in the presence of a suitable solvent.

The resin is mixed with the solvent in a beaker to form a slurry and then the slurry is poured into the column. Demineralised water should be used; that is, water free of any ions. The column needs to be cleared of air bubbles. This is achieved by an upward flow of water. A water supply is attached to the bottom of the column, the tap is opened and the water is forced up into

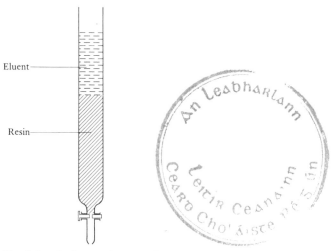

Fig. 4.2   An ion exchange column

the column from the bottom. When no further air is displaced, the resin is allowed to settle and the liquid is run out until it is just above the top of the resin. The ion mixture is added to the column using a minimum of solvent. This mixture is now allowed to run into the resin and be absorbed by it. The surface of the resin is then covered (for example, with a small filter paper or cotton wool) to prevent any disturbance of the beads. The apparatus is then topped up with the appropriate eluent (the solvent used for the separation). The solvent runs through the resin slowly. The rate of elution is important so that equilibrium may be established at each level of beads. If the resin is required in a form other than that in which it is supplied by the manufacturer (e.g. an acid form rather than the sodium form), it is washed with a solution of an appropriate ion and then with demineralised water before charging with the mixture. It has been shown above that the exchange potential for sodium ions is greater than that for hydrogen ions in dilute solutions. The reverse process, as required here, can be achieved by using concentrated acid.

The volume occupied by a resin after the preparation of the column is called the 'bed volume' (BV). This is the volume occupied by the beads and the void spaces between them. Volumes of eluent applied are referred to in terms of the bed volumes, and the rates of liquid flow in bed volumes per minute (e.g. 0.20 BV min^{-1}).

The **exchange capacity** of a resin is measured in terms of the number of moles of electrical charge per cm^3 of wet resin. Typically, for a cation exchange resin, the exchange capacity is 2×10^{-3} mol cm^{-3} in the hydrogen form. The nature of the counter ion is usually quoted because the swollen volume of the resin is affected by this. Experiment 4.2 describes the determination of the capacity of a cation exchanger. An acidic resin is washed with sodium chloride solution to displace the hydrogen ions. The eluent is titrated against standard hydroxide solution to determine the number of moles of H$^+$ displaced.

When an ion exchange resin has been used, it can be regenerated into its original form. Regeneration is performed using approximately 1 molar solution of the displaced ion. For strongly acidic or basic resins the quantity of ions used should be about four times the capacity of the column (see below). Weakly acidic or basic resins require an amount double the capacity.

For example, consider the regeneration of 150 cm^3 strongly acidic resin of capacity 2×10^{-3} mol cm^{-3} to its acid form with 1 M HCl. Since it is a strongly acidic resin, a fourfold quantity is required. So, the volume of

$$\text{acid required is equal to } \frac{4 \times \text{ capacity (mol cm}^{-3}) \times \text{ BV (cm}^3)}{\text{molarity (mol dm}^{-3})}$$

$$= 4 \times 2 \times 10^{-3} \times 150 / 1$$
$$= 1.2 \text{ dm}^3$$

Laboratory applications of ion exchange

Often ion exchange resins are used in columns as described previously. This is not always necessary or convenient; they may be stirred with the reactant solution in a flask or other suitable reaction vessel (see, for example, experiment 4.5). Paper can be used as a medium for ion exchange chromatography. Either a piece of chromatography paper (p. 95) can be impregnated with an ion exchanger by dipping it into a solution of the substance, or the paper may be prepared from a cellulose derivative instead of cellulose itself. A cation exchanger is formed, for example, by the use of carboxymethylcellulose or cellulose citrate; aminoethylcellulose would produce a paper capable of exchanging anions. The paper is used in the manner described for paper chromatography (p. 96).

(a) Separation of ions

Ions can be separated if they have different distribution coefficients. When one of the pair of ions has been eluted, the solvent (or its pH) may be changed to reduce the distribution coefficient of the other ion and increase its rate of elution. For example, a chloride–bromide mixture can be separated by 0.3 M sodium nitrate on a strongly basic resin (nitrate form). After the chloride has been eluted, the eluent is changed to 0.6 M sodium nitrate. Every 10 cm^3 of eluent is titrated against silver nitrate solution to determine the amount of halide displaced. Figure 4.3 gives a typical graph of the titre against total volume eluted (see experiment 4.3). This technique has been successfully employed to separate the lanthanides and actinides from mixtures of the ions (Fig. 4.4).

(b) Demineralisation

Demineralisation is the removal of ions from a solution. This is achieved in two stages – replacement of metal cations by hydrogen ions:

$$\text{(R)}{-}\text{H} + \text{M}^+\text{(aq)} \rightleftharpoons \text{(R)}{-}\text{M} + \text{H}^+\text{(aq)}$$

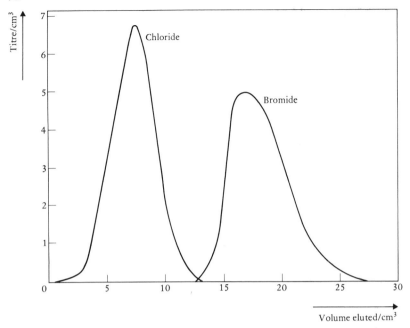

Fig. 4.3 Titration curve for the elution of halide ions from an ion exchange resin

and of anions by hydroxide ions:

$\textcircled{R}$—OH + A⁻(aq) $\rightleftharpoons$ $\textcircled{R}$—A + OH⁻(aq)

The H⁺ and OH⁻ ions then react to give neutral water molecules:

H⁺(aq) + OH⁻(aq) $\rightleftharpoons$ H₂O

(c) Preparations

Weak acids are often difficult to prepare, frequently being decomposed by the concentrated acids which would be used in their preparation. On other occasions, the method of preparation is complex. Ion exchange resins can be used to prepare them from their salts. Typical examples include

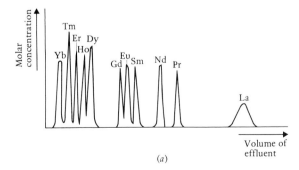

(a)

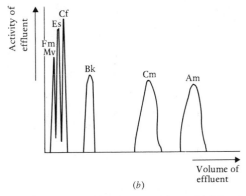

Fig. 4.4 Ion exchange separation of (*a*) lanthanides and (*b*) actinides

H_4[Fe(CN)$_6$], hexacyanoferric(II) acid, and $HMnO_4$, manganic(VII) acid. Similarly, hydroxides can be prepared using anion exchangers.

(*d*) Catalysis

An acidic resin contains hydrogen ions. Acid catalysed reactions may, therefore, be performed using acidic resins. The inversion of sucrose, the hydrolysis of esters, and the dehydration of alcohols are examples of this process. One advantage of using resins is that they can be filtered off from the reaction mixture and so do not contaminate the products.

Basic resins can catalyse the hydrolysis of nitriles to amides and the formation of aldols and cyanohydrins.

(*e*) Analysis

It is difficult to estimate some ions directly; an ion exchange resin can be used to produce a suitable alternative ion. For example, a sodium salt eluted through an acidic resin will be converted to an acid of equivalent molarity. The acid content of the eluent can then be determined by titrimetry.

Industrial applications of ion exchange

Industrially, similar applications are used. Water often has to be softened (that is, demineralised) before use in boilers. The sugar syrup extracted from beet is treated with an ion exchange resin to remove inorganic salts before crystallisation of the sugar. Many wines are supersaturated with potassium hydrogen tartrate which precipitates out on standing. This can be overcome by passing the wine through a sodium ion resin to produce the more soluble sodium salt. Acids used in metal treatment plants (e.g. phosphoric acid for steel and chromic acid for anodising) can be regenerated using an acidic resin. Uranium is concentrated from its ores as a sulphate complex on an anionic resin. Ion exchange resins can be used in medicine and medical research. For example, a condition called hyperkalaemia is due to excessively high concentrations of potassium (Latin, *kalium*) in the blood following kidney failure, and leads to headaches, loss of appetite, vomiting and, ultimately, heart failure. Treatment includes the use of sodium or calcium ion exchange resins; the sodium ion resin is sometimes unsuitable as this can lead to a build up of sodium in the blood with the consequent formation of fluid in the lungs.

Chromatography

Chromatography is a means of separating mixtures by distribution of the components between a stationary phase and a mobile phase. A number of applications of this principle have been developed and are described in Table 4.2. Figure 4.5 shows the principle of separation; the two components

Table 4.2   Types of chromatography

Description	Stationary phase	Mobile phase
Paper	Adsorbed water on paper	Liquid
Thin layer	Adsorbed water on powder	Liquid
Column	Adsorbed water on powder	Liquid
Dry column	Surface of powder	Liquid
Gas–liquid	Non-volatile solvent on powder	Inert gas
Gas–solid	Surface of powder	Inert gas
Molecular sieve	Silicate	Liquid
Gel permeation	Polysaccharide	Liquid
Ion exchange	Ionic sites on resin	Aqueous agent

(represented by dots and crosses) have differing distribution ratios between the two phases. As the mobile phase flows through the column of stationary phase, the components are gradually separated. The component represented by the crosses is held by the stationary phase to a greater extent than is the component represented by the dots. After the seven stages illustrated, the component shown as dots is more advanced than the component represented by the crosses.

Chromatographic separations can be achieved by the adsorption or partition mechanisms. In **adsorption chromatography** the components of a mixture are separated by their different adsorption/desorption behaviours in the presence of a mobile solvent. The substances have varying interactions with the surface of the solid particles (due to hydrogen bonding or weak electrostatic forces) giving adsorption and differing affinities with the solvent. Adsorption chromatography may be classified as liquid/solid or gas/solid chromatography, depending on the nature of the mobile phase used. Weak adsorbents include starch and talc; magnesia, alumina and silica gel are relatively strong adsorbents.

Partition chromatography, which is used more extensively and is described in detail in the techniques that follow, involves the distribution of the components of a mixture between an adsorbed liquid film (the stationary phase) and a mobile liquid. A commonly adsorbed liquid is water, which is present in paper, for example.

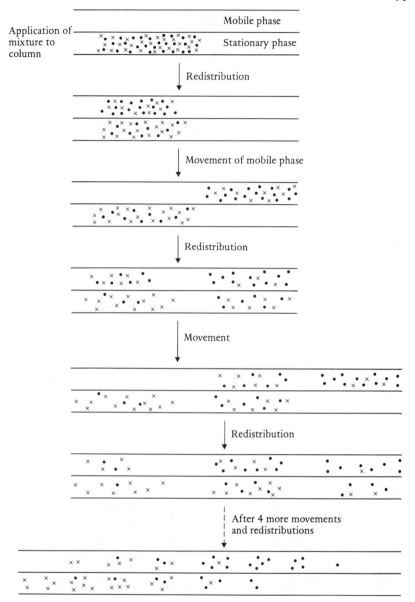

Fig. 4.5 Principles of chromatographic separation (distribution ratios, mobile phase/stationary phase; · 2/1 and x 1/2)

(a) Paper chromatography

An advantage of this technique is its speed, but it is only suitable as a means of qualitative analysis. The mobile phase, called the eluent, is allowed to rise up the paper (ascending paper chromatography) or soak down the paper (descending paper chromatography) as shown in Fig. 4.6. The mixture is

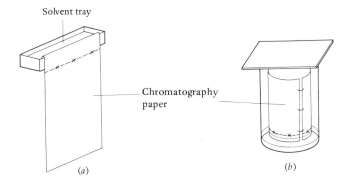

Solvent tray

Chromatography paper

(a)

(b)

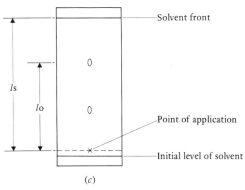

Solvent front

l_s

l_o

0

0

Point of application

Initial level of solvent

(c)

Fig. 4.6   Paper chromatography

applied to the paper at a spot beyond the initial solvent front. The spot should be as small as possible; large spots give very diffuse spots after separation. The solvent moves across the paper, separating the components. The container is covered so that the atmosphere is saturated with solvent. If it is uncovered, the solvent evaporates from the paper. After the solvent has traversed most of the paper, the paper is removed from its container and dried. The final position of the solvent front is marked. Each substance moves a characteristic distance relative to the solvent front. This distance is known as the R_f value and is defined as $\dfrac{\text{distance moved by component}}{\text{distance moved by the solvent front}}$. The R_f value is dependent on the substance, solvent, paper and temperature. The components are most reliably identified by running reference samples under the same conditions (Fig. 4.7). The reference samples used are of the

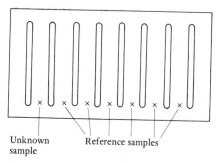

Fig. 4.7   Use of split-papers for reference samples

possible components of the mixture. For example, a solution is suspected to contain the ions of calcium, strontium and barium. The three reference solutions contain one of these ions each. So, a chromatogram is developed using a sample of the mixture, a sample of a calcium ion solution, a sample of a strontium ion and finally a sample containing barium ion. When development is complete, the positions of the spots from the reference samples are compared with any spots formed on elution of the mixture. If spots occur for the unknown in positions corresponding to the calcium and barium ions in the reference samples, but not corresponding to strontium, probably only the former two ions are present in the mixture.

An alternative term to R_f can be used to represent the progress of the substance compared to the progress of a reference substance; this is the R_x term. R_x is defined as $\dfrac{\text{the migration distance of the substance}}{\text{the migration distance of a reference}}$. The R_f value is always less than unity; R_x values are greater than zero.

If the substances being separated are coloured, then detection of the positions of the components is easily achieved. In the case of colourless substances, they have to be detected by conversion to a coloured derivative or otherwise detectable substance. For example, amino acids treated with ninhydrin solution give purple spots on warming; carbonyl compounds give yellow spots when treated with 2,4-dinitrophenylhydrazine; fluorescent substances can be detected by exposure to ultraviolet radiation. The R_f values of the substances are not affected by the detecting agent, the transport being achieved prior to detection.

If a group of substances is inadequately separated by one solvent, then the paper may be turned through $90°$ and a second solvent used (Fig. 4.8); this is two-way paper chromatography.

(b) Column chromatography

Column chromatography is used for the quantitative separation of mixtures. A column of powder (usually alumina, Al_2O_3, or silica gel) is set up by preparing a slurry of the powder in the solvent. This is transferred to a glass tube with stirring, to exclude air bubbles, and is then allowed to settle.

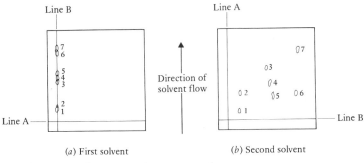

(a) First solvent   (b) Second solvent

Fig. 4.8   Use of two-way chromatography

A good separation often requires about 1 kg of powder for each gram of mixture used. The solvent is run down to just above the powder level and the mixture is carefully applied. The mixture must be introduced carefully so that the powder is not disturbed; failure to observe this precaution will result in a poor separation of the components. The mixture is run on to the column, which is then covered with cotton wool, for example, and the eluent is added and allowed to penetrate the column slowly. As one component is eluted, the solvent may be changed to separate out other components or to speed up the elution of the other slow moving substances. If the most suitable solvent is not known in advance, it is normal to start elution with a solvent of low polarity and gradually increase the polarity. For example, the sequence hexane, methylbenzene, ethoxyethane, ethyl ethanoate, methanol might be used. The polarity can be controlled by mixing solvents, e.g. methylbenzene, then 1 per cent ethoxyethane in methylbenzene, then 10 per cent, 50 per cent and finally 100 per cent ethoxyethane.

(c) Thin layer chromatography (TLC)

Thin layer chromatography has the versatility of column chromatography (in the nature of supporting materials available) and is even faster than paper chromatography. The solid phase is a powder supported on to glass, metal or a 'plastic' substance. Microscope slides are excellent supports; commercial plates are prepared on metal and polymer foils and can be cut to size.

The plates are coated with a thin, uniform slurry of the absorbent material (cellulose, alumina, silica gel, ion exchange resin, etc.). As the solvent evaporates, the powder adheres to the support. If there is poor adherence, a binding agent such as calcium sulphate may be added.

Small spots of the substance under investigation are applied to the plates and the chromatogram is developed by the ascending method. Chromatograms on glass plates are difficult to preserve and include in a practical report, but they may be included in the documentation by transfer to clear adhesive tape (e.g. Sellotape). A piece of tape, larger than the plate, is laid across the dried chromatogram. When the tape is lifted off, the chromatogram is adhering to it and may be transferred to the laboratory report.

(d) Gas chromatography

This technique achieves separation by the distribution of a substance between a non-volatile oil or grease (held as a stationary phase by adsorption on an inert solid) and an inert gas; it depends on the solubility of the material in the liquid and its volatility at the temperature of the separation. The basic arrangement for the process is as shown (Fig. 4.9). The column is heated in an

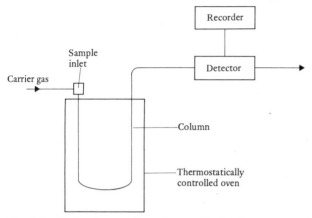

Fig. 4.9 Basic construction of a gas—liquid chromatography apparatus

oven. The greater the column temperature, the greater is the tendency for the solute to favour the gaseous phase and so the lower is its retention time. Temperature control must be very precise if results are to be reproducible. For reproducible results a constant pressure of the inert gas is also required. The gas must not affect the detector signal and usually must be pure and dry. Nitrogen is a suitable example.

One of the easiest methods for the introduction of a small sample of the mixture to be analysed is to inject the sample from a micro-syringe through a rubber serum cap. The inlet is often heated by a small electric coil to prevent condensation of mixtures of low volatility.

The nature of the column is determined by the materials to be analysed. The solid support is an inert material such as Celite (a form of silica, SiO_2) or glass beads, the tube being constructed of metal (glass may also be used). The solid is coated with a non-volatile oil or grease as solvent (e.g. silicone oil or polyethylene glycol). The oil must act as a solvent for the components of a mixture; it must be chemically inert to the gas and mixture and must be thermally stable under the operating conditions. Hydrocarbons can be conveniently separated on silicone oil or dinonyl phthalate; polyethylene glycol is more suitable for polar substances such as fatty acids. Common dimensions for the columns are 2–5 m long and 2–10 mm diameter.

If the column is to achieve separation by adsorption on a solid (gas–solid chromatography) rather than by solution in the oil (gas–liquid chromatography), suitable packing materials are alumina, silica gel, activated charcoal and molecular sieves. The packing must be uniform in order to ensure sufficient separation.

There are a wide range of detectors available. The **flame ionisation detector** operates on the principle that the electrical conductivity of a hydrogen flame is increased when easily ionisable materials enter the flame. Normally the carrier gas is nitrogen and hydrogen is introduced at the outlet to produce the flame. A potential of about 200 V is sufficient to produce a small current (Fig. 4.10). The current is detected using an amplifier. While

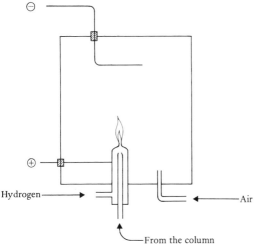

Fig. 4.10   Flame ionisation detector

effective with the easily ionisable organic molecules, it is less effective with the inorganic compounds which cannot be ionised so readily or even not at all.

A **katharometer** is based on the variation in the thermal conductivity of the carrier gas when the components of the mixture are eluted. The katharometer consists of two identical cells which measure the thermal conductivity of the carrier gas before and after entering the column. The cell (Fig. 4.11)

Fig. 4.11   Part of a katharometer

consists of an inert electric wire (e.g. tungsten) heated electrically. If only pure nitrogen is present in each, the resistance of each wire (detected by a Wheatstone bridge circuit) is identical and is, therefore, balanced. The presence of an impurity in the gas causes a change in resistance and so the bridge is unbalanced; the galvanometer shows the deviation by a deflection. The products are not destroyed by this method and so can be collected in preparative work.

Other techniques include the **flame temperature detector** (the combustion of organic compounds is generally exothermic) and the **argon ionisation**

detector (argon is activated by beta-radiation and causes the ionisation of organic molecules). A simple experiment showing how the components of a mixture can be determined chemically (by titration) is given in experiment 4.14.

The changes indicated by the detectors are recorded on a chart recorder. An idealised record of a separation by GLC would be as shown in Fig. 4.12.

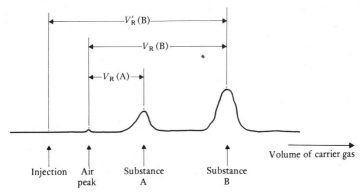

Fig. 4.12 Idealised record of a GLC separation

The air peak is formed when the mixture is injected. The volume of gas eluted from the time of injection to the time of elution of the substance under investigation is called the **retention volume,** V_R'. The volume eluted from the air peak to the elution of the substance is the **adjusted retention volume,** V_R. The difference between these values $(V_R' - V_R)$ is the 'free gas volume'. When air is not detectable (as in a flame ionisation detector) the retention volume can be measured from the time of injection of the mixture or by the inclusion of a suitable non-sorbed substance in the mixture. This will give a peak in a similar position to the air peak.

The distances $V_R(A)$ and $V_R(B)$ are known as the adjusted retention volumes of A and B respectively, that is, the volume of gas required to elute the components from that column under the given pressure and temperature conditions. The areas under the peaks are proportional to the quantities of each component present. For the flame ionisation detector system, the amount of each component eluted can be determined by the use of an internal standard. For example, if the amount of A is known (Fig. 4.12), the quantity of B can be determined by comparison of the peak areas:

$$\frac{\text{amount of A}}{\text{amount of B}} = \frac{\text{area of peak A}}{\text{area of peak B}}$$

For other detector systems, this comparison of peak areas does not necessarily give the concentration ratios. These ratios are affected by a number of factors including, for example, the thermal conductivity of the material. In these cases a quantitative analysis is only possible by injecting a known quantity of the substance into the column under the same experimental conditions as those used with the unknown quantity.

(e) Recent developments in chromatographic methods

More efficient and rapid column chromatography is achieved using **high performance liquid chromatography.** The increase in speed of separation is achieved by pumping the solvent through the column under pressure. A more efficient separation results from the use of smaller particles of column material and better packing techniques.

Affinity chromatography has been developed particularly for use with biologically active molecules. The resin support is treated to produce active groups on it. For example, cellulose can be treated to produce a bromo derivative:

$$Cel-OH + Br\text{-}COCH_2\,Br \rightarrow Cel-O-COCH_2\,Br + HBr$$

If a solution of proteins is added to the column, a reaction occurs, binding the protein to the resin:

$$Cel-O\text{-}COCH_2\,Br + H_2\,N\text{-protein} \rightarrow Cel-O\text{-}COCH_2\,NH\text{-protein} + HBr$$

Any substances lacking affinity for the functional group pass through the column unretarded. The proteins can be detached from the resin by elution with acid. The protein is, therefore, separated from contaminants. A careful control of the pH can cause the selective detachment of proteins present in a mixture and so facilitate their separation from each other.

Gel permeation chromatography and **molecular sieve chromatography** achieve separation through the filtration effect of the pores of these materials. The technique has particular value in the separation of molecules of similar polarity but of different size, traditional methods of chromatography being less effective in these circumstances.

(f) Electrophoresis

Electrophoresis is an extension of the chromatographic technique in which an electrical potential is used as the 'driving force'. A piece of chromatography paper is wetted with a buffer solution (so that it conducts electricity) and the mixture is applied at the centre (Fig. 4.13). The ends of the paper dip into

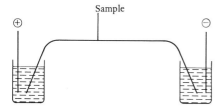

Fig. 4.13   Electrophoretic separation

separate reservoirs of the buffer and an electrical potential is applied. Cations migrate to the cathode, anions to the anode. The rate of migration is determined by the charge, mass, potential, pH, etc. Separation may be performed on a thin layer film. The pH of the buffer is particularly important with substances such as amino acids which can take up positive or negative charges:

$$\underset{\substack{\text{H}_2\text{N}-\text{C}-\text{COO}^- \\ |\\ \text{H}}}{\overset{\substack{\text{R} \\ |}}{}} \quad \overset{\text{OH}^-}{\rightleftharpoons} \quad \underset{\substack{\text{H}_2\text{N}-\text{C}-\text{COOH} \\ |\\ \text{H}}}{\overset{\substack{\text{R} \\ |}}{}} \quad \overset{\text{H}^+}{\rightleftharpoons} \quad \underset{\substack{\text{H}_3{}^+\text{N}-\text{C}-\text{COOH} \\ |\\ \text{H}}}{\overset{\substack{\text{R} \\ |}}{}}$$

$+ H_2O$

Anion at high pH   Cation at low pH

Solvent extraction

(a) Soxhlet extraction

It is sometimes necessary to perform repeated extractions of a mixture with a hot solvent in order to achieve an adequate recovery of the required material. This is achieved using a Soxhlet extractor (Fig. 4.14). The material

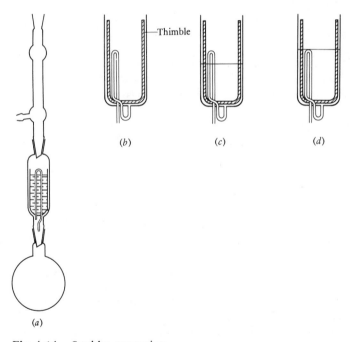

(a)

Fig. 4.14   Soxhlet extraction

to be extracted is placed in the porous thimble and the solvent is heated in the flask. The hot vapours are condensed into the thimble where the liquid dissolves some of the solute (Fig. 4.14c). When the liquid level reaches the top of the narrow tube, it is siphoned off (Fig. 4.14d) into the flask. The solvent is recycled. After the extraction is complete, the solvent is distilled off to isolate the solute.

(b) Simple extraction

The principle of partition between two liquids (p. 44) is applied in extractions into ethoxyethane ('ether'). In many organic preparations, the required product dissolves in a solvent from which it cannot be conveniently extracted. Ether is added to the solution and the product is distributed between the two liquids. As the following calculation shows, it is more efficient to extract with several small portions of ether than to use one large bulk.

A preparation yielded 1 g succinic acid in 100 cm^3 of water. The partition coefficient for succinic acid, $K_{ether/water}$, is 5. Succinic acid is unassociated in both solvents. Consider first the use of a single 100 cm^3 portion of ether. Let x g pass into the ether layer.

$$K = \frac{\text{Concentration of acid in ether}}{\text{Concentration of acid in water}}$$

$$5 = \frac{x/100}{(1-x)/100} = \frac{x}{(1-x)}$$

Hence, $x = 0.83$

So, 0.83 g of the succinic acid is extracted by the ether.

Now consider the ether added in two separate 50 cm^3 portions. Let y_1 g be extracted in the first portion and y_2 g be extracted in the second portion.

$$5 = \frac{y_1/50}{(1-y_1)/100} = \frac{2y_1}{(1-y_1)}$$

$$y_1 = 0.71$$

0.29 g are left in the water layer.

$$5 = \frac{y_2/50}{(0.29-y_2)/100} = \frac{2y_2}{(0.29-y_2)}$$

$$y_2 = 0.21$$

So, the total transferred in the second technique is 0.92 g, in contrast to 0.83 g in the first method. It is, therefore, more beneficial to extract a solute from a solution using two portions of solvent rather than one portion of the same total volume.

Often, in solvent extraction, one solvent is aqueous and the other solvent is organic. If one component of a mixture to be separated is basic, its solubility in the aqueous phase is enhanced by the use of an acid. For example, if an amine is present, it can be extracted using hydrochloric acid:

$$RNH_2 + HCl(aq) \rightarrow RNH_3{}^+ Cl^-(aq)$$

The amine can be released from the aqueous solution by addition of sodium hydroxide solution and extracted by ethoxyethane.

$$RNH_3{}^+ Cl^-(aq) + Na^+ OH^-(aq) \rightarrow RNH_2 + Na^+ Cl^-(aq) + H_2O$$

Similarly, an acid can be extracted using a basic medium and a carbonyl can be extracted as its hydrogen sulphite complex:

$$RCHO + Na^+ HSO_3^-(aq) \rightarrow RCH(OH)SO_3^- Na^+(aq)$$

A mixture of organic acids with differing pK_a values may be separated by altering the pH so that one is dissociated while the other is not.

If a mixture of polar substances in water is to be separated, then extraction by a solvent of low polarity can be achieved if the charge on one of the ions can be reduced. This is conveniently done by a process of complexation. One complexing species of this type is 8-hydroxyquinoline (Fig. 4.15). This reacts with some metal ions (usually in weak acid or neutral solutions) to form covalent compounds. The metal ion becomes part of a covalent complex and is now soluble in an organic solvent such as trichloromethane. Other complexing agents are penta-2,4-dione, dithizone, dimethylglyoxime, etc. Groups like these that coordinate to a metal ion are called **ligands.** If a ligand bonds to the metal ion through two or more atoms (Fig. 4.15), it is known as a **chelating** ligand.

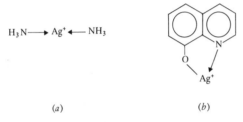

Fig. 4.15   Coordination of (a) a simple ligand (ammonia) and (b) a chelating ligand (8-hydroxyquinoline) to a silver ion

(c) Multiple extractions

It has been shown that a two-stage extraction is more effective than a single-stage extraction (p. 104). This can be extended to give a multistage, or countercurrent, extraction system. The effect can be illustrated as follows.

A mixture of two solutes A and B is dissolved in an aqueous medium and the solution is extracted with an organic solvent. Let A have equal solubilities in each solvent and B be four times as soluble in the organic layer as in the aqueous layer. After the first extraction, half of A and four-fifths of B are in the organic layer. This layer is removed and extracted with an equal volume of the aqueous layer; the aqueous layer remaining in the first container is extracted with another portion of the organic solvent. This process is repeated by transferring the organic layers — from container 2 to 3 and from 1 to 2; a fresh supply of organic liquid is added to 1 (see Fig. 4.16).

Table 4.3 gives the percentages of transference at different stages. The results of this are shown graphically in Fig. 4.17. It will be seen that the separation is complete after 100 transfers. A number of commercial systems have been developed to do this process automatically.

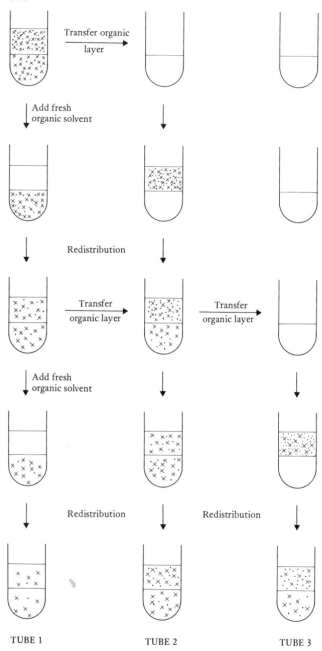

Fig. 4.16 Principle of multistage extraction

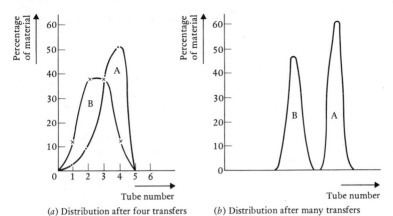

Fig. 4.17   Percentage transference in multistage separations

Table 4.3   Percentages of transference in multistage extractions

| | | Tube: percentages of original material | | | |
		1	2	3	4
Initially	A*	100%			
	B	100%			
After one transfer	A	20%	80%		
	B	50%	50%		
After two transfers	A	4%	32%	64%	
	B	25%	50%	25%	
After three transfers	A	0.8%	9.6%	38.4%	51.4%
	B	12.5%	37.5%	37.5%	12.5%

*A corresponds to . and B to x in Fig. 4.16.

The process of partition chromatography (adsorbed water–eluent partition) provides a means of separation involving a very large number of transfers and so is a very efficient means of separation.

Distillation
(a) Simple distillation

A solution of a solute in a solvent boils at a temperature higher than that of the pure solvent (see *Fundamentals of Chemistry*, p. 80). The boiling point of the solution is dependent on the molar concentration. The relationship between the boiling point of a solution and its concentration is given by

$$\Delta T = \frac{1\,000 \cdot K \cdot w}{W \cdot m}$$

where ΔT is the elevation of the boiling point (above that of the pure solvent),

K is the ebullioscopic constant (dependent on the nature of the solvent),

w, W are the masses (in kg) of solute and solvent respectively,

and m is the relative molecular mass of the solute. This relationship only holds for dilute solutions of the solute. At higher concentrations, it is necessary to make allowance for interactions between the solute and solvent species.

Dissociation and association give abnormal results because the number of moles of dissolved particles varies. The boiling point of a liquid is also affected by atmospheric pressure (Fig. 4.18).

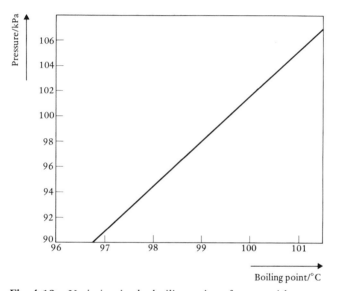

Fig. 4.18   Variation in the boiling point of water with pressure

If the solute is non-volatile, and so is not displaced from solution during the boiling process, the temperature of vapour is that of the vapour of the pure solvent boiling at the same pressure. For example, if water boils at 373 K, then a solution of sodium chloride in water (boiling at, say, 374 K) will give steam at 373 K.

A solute extracted into a solvent (e.g. succinic acid into ether in the example on p. 104) can be recovered by simple distillation of the solvent (Fig. 4.19). The basic apparatus is arranged so that the thermometer records the temperature of the vapour distilling into the condenser. This is important if two components are to be distilled off separately; the thermometer reading indicates when the first component has been completely removed.

If the solvent is flammable, no naked flames should be used; a hot water bath is a good alternative in the case of, say, ethoxyethane solvent. Very volatile liquids such as ethoxyethane should be collected in an ice bath to

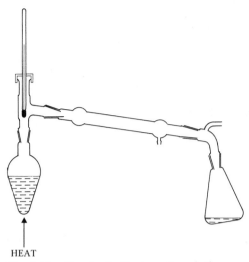

HEAT

Fig. 4.19   Simple distillation of a solution

ensure maximum possible condensation; the fumes should also be directed to some safe extraction system. Be careful not to fit the apparatus without an outlet system for the uncondensed vapours; the air in the apparatus together with the vapour will expand during the distillation and blow the equipment apart as the pressure builds up!

If the hot vapours entering the condenser are above 410 K (140 °C), an air condenser should be used. Above this temperature, a water condenser is liable to crack due to the temperature gradient across the glass (cold water on one side; hot vapours on the other).

All distillations should include the addition of fragments of unglazed porcelain (or 'anti-bumping granules'). These provide nuclei for the formation of bubbles of vapour and so the boiling proceeds in a steady manner. Exclusion of these particles is likely to result in superheating, the liquid boiling suddenly and jumping in the flask, usually passing straight into the condenser before complete vaporisation has occurred. This can result in a breakage of the flask, a loss of product, the contamination of the pure distillate and injury to the experimenter.

(b) Fractional distillation of miscible liquids

For miscible liquids, the vapour pressure varies with the composition of the mixture. Raoult has formulated a law which equates the total vapour pressure to the sum of the products of the mole fractions and partial pressures of the components (Fig. 4.20). So,

$$p = x_1 \cdot p_1 + x_2 \cdot p_2$$

where p = total vapour pressure, p_1 and p_2 are the pressures of components 1 and 2, x_1 and x_2 are the mole fractions of the respective components. Figure 4.21 gives the temperature–composition diagram for a completely miscible system (methylbenzene–benzene). If a liquid corresponding to the

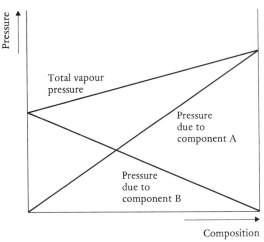

Fig. 4.20   Graph of Raoult's law

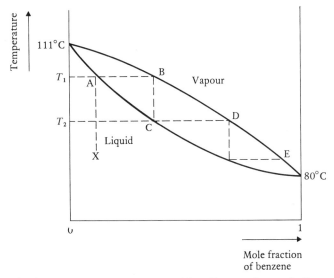

Fig. 4.21   Temperature–composition diagram for methylbenzene–benzene mixture

point X (Fig. 4.21) is boiled, it boils at point A. The vapour at that tempera-ture (T_1) will have composition B. Since the vapour B is richer in benzene than the liquid A, the liquid must now become richer in methylbenzene. Its boiling point will, therefore, increase. As more vapour is produced from this liquid, the boiling point gradually increases to 111 °C. The vapour B rises from the flask and is condensed by the air to a liquid of identical composition (C). This liquid C has a lower boiling point (T_2) than the original liquid A and is boiled by more hot vapours. The vapour D produced

from this sample (C) is richer in benzene than was vapour B. Since it has
a lower boiling point than A/B, it will rise higher up the condenser before
it is cooled sufficiently to condense. It will then reboil to give more vapour
E. The vapours rise higher as they become richer in benzene. Eventually
pure benzene results with boiling point 80 °C. So, the vapour produced is
benzene and the liquid remaining is methylbenzene. This is the principle
applied in **fractional distillation.**

Various columns have been designed to facilitate this means of
separation (Fig. 4.22). The basic characteristics of these columns are a
large surface area and efficient mixing of the vapour and condensed liquids.

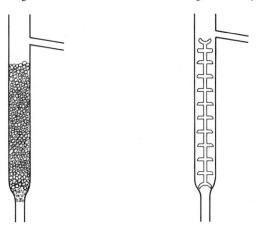

Fig. 4.22   Fractional distillation columns

The packing materials can be beads, pieces of tubing, rings, etc. Figure 4.23
illustrates the procedure. The mixture of composition A boils and the vapour

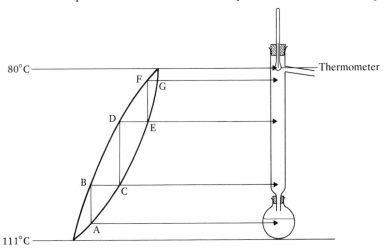

Fig. 4.23   Principles of fractional distillation

enters the column. B rises up the column until it is condensed by the cooling effect of the surroundings. The condensate, C, runs down the column and is heated by fresh vapours rising up the column and is reboiled by them. The new vapour, D, rises further up the column since it needs to be cooled down further before it is condensed to a liquid, E. E is reboiled by more rising vapour, so giving vapour F and so on. The process of condensation and reboiling continues in the column until only the lower boiling component is left to run from the column and the higher boiling component remains in the flask. The rate of heating must be controlled so that the lower boiling point (in this case 80 °C) is not exceeded at the top of the column. If it is exceeded, then vapour originating from a mixture of higher boiling point (e.g. point G) can also leave the column, thereby contaminating the product. When the lower boiling component has been completely removed, no further liquid will be condensed until the higher boiling component is boiled off.

Some liquids do not obey Raoult's law (Fig. 4.24); if the molecules of the mixture have less affinity for each other than do the molecules of the pure components, they vaporise more easily than predicted by Raoult. So, the vapour pressure is higher and the boiling point of the mixture is lower than expected. A maximum results in the vapour pressure–composition curve and this is described as a positive deviation from Raoult's law. An

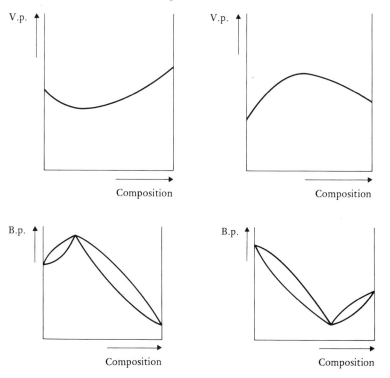

Fig. 4.24 Abnormal Raoult's law results

example of this is an ethanol–benzene mixture; a maximum occurs for a 95.6 per cent ethanol content. Ethanol and water also have a positive deviation.

When the molecules of the liquid mixture have stronger attractions for each other than that existing in the pure components, then a negative deviation results. Typical examples include nitric acid–water and trichloro-methane–methyl methanoate mixtures.

In these two deviation situations, a complete separation is not possible. In the case of a negative deviation, the boiling point gradually rises to a maximum value; the boiling point and composition of the mixture remain constant at this point and the mixture is called a **constant boiling mixture** (C.B.M.). These mixtures are known as **azeotropes.** For mixtures involving a positive deviation, the constant boiling mixture has a minimum boiling point. The complete separation of the components by distillation is difficult. One common method is to use a chemical which reacts with one component only. For example, if quicklime (CaO) is added to the aqueous ethanol azeotrope, the ethanol can be distilled off, the quicklime absorbing the water.

(c) Steam distillation

A pure liquid boils when the vapour pressure is equal to the atmospheric pressure. For two immiscible liquids, the total vapour pressure is equal to the sum of the individual vapour pressures. Consequently, the mixture boils at a lower temperature than the boiling points of the individual components. For example, aniline (b.p. 457 K) has a vapour pressure of 7 kPa at 371 K; water (b.p. 373 K) has a vapour pressure of 94 kPa at the same temperature. The total vapour pressure of aniline and water at 371 K is 101 kPa and so the mixture boils at this temperature. This is the principle used in **steam distillation.**

Aniline is produced in a reaction resulting in a mixture with inorganic salts. It can be separated by steam distillation (Fig. 4.25). This allows the

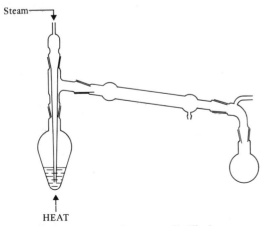

Steam

HEAT

Fig. 4.25 Apparatus for steam distillation

separation of aniline at a temperature well below its boiling point and is useful when the compound is unstable at temperatures approaching its boiling point. In this example, the aniline is extracted by the steam at 371 K. The condensate (aniline and water) separates as two layers; the aniline is removed, dried and redistilled to remove any remaining impurities.

A further example of the application of steam distillation is in the separation of the isomers 2- and 4-nitrophenols.

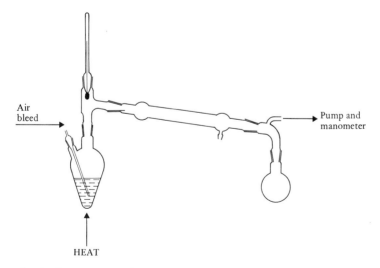

In 2-nitrophenol, hydrogen bonding is intramolecular, but it is inter-molecular in 4-nitrophenol. The 4-nitrophenol has a high melting point relative to the 2-nitrophenol (114 °C and 46 °C respectively) and is less volatile than the latter compound. As a result, a solution of the two isomers (as produced in the nitration of phenol) can be steam distilled, the more volatile isomer, 2-nitrophenol, distilling off with the steam.

(d) Low pressure distillation

Some compounds undergo decomposition at or below their boiling points at standard pressure. These substances can be distilled at reduced pressures. Figure 4.26 illustrates a typical arrangement of the apparatus for such a

Fig. 4.26 Apparatus for distillation at low pressure

system. A manometer is necessary so that the boiling point can be related to the pressure (see Fig. 4.18).

Anti-bumping granules are inadequate to prevent superheating at low pressure and so an 'air bleed' is used. Nitrogen replaces the air if the compound is air-sensitive.

Industrial applications

Distillation is used in industry for a wide variety of processes. Fractional distillation is used in the petroleum oil industry for the separation of the hydrocarbon mixtures. Because of the large number of hydrocarbons and the closeness of their boiling points, the separation is not into specific compounds, but into groups of substances with similar physical and chemical properties.

The fractionating column is constructed from a series of trays (Fig. 4.27) containing perforations covered by 'bubble caps'. The trays are linked to the

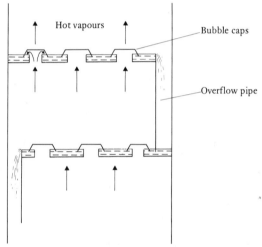

Fig. 4.27 Diagrammatic representation of the industrial fractionating column

outlets through pipes as shown. The vapour from the heated oil rises up the column and condenses at a height related to its boiling point. The higher boiling liquids are condensed out at the lower part of the column; the more volatile compounds condense higher up. The liquid collects on the trays. Further hot vapour rises through the bubble caps and passes through the condensed liquid, so displacing any more volatile components. As the condensate fills the tray, it overflows to the one below, which is at a higher temperature; the more volatile components are boiled out again and rise further up the column. The less volatile material is tapped off. In this manner a number of useful fractions are obtained (Table 4.4; see *Fundamentals of Chemistry*, p. 197, for further details).

Table 4.4 Distillation of oil

Boiling temperature of fraction ($^\circ$C)	Fraction name	Hydrocarbon content	Uses
<20	Gas	$C_1 - C_4$	Heating
20−60	Light petroleum	$C_5 - C_6$	Solvent
60−100	Ligroin	$C_6 - C_7$	Solvent
40−205	Petrol	$C_5 - C_{12}$	Fuel
175−325	Kerosene	$C_{12} - C_{18}$	Heating fuel
275−400	Gas oil	$C_{12} - C_{25}$	Diesel fuel
Non-volatile liquid	Lubricating oil		Lubrication
Residue	Asphalt or bitumen		Road tar

Low pressure distillation is particularly important for the concentration of foodstuffs. High temperature distillation is destructive to vital components of food (e.g. proteins), and so the excess water must be removed at a lower temperature (and therefore lower pressure).

Summary

At the conclusion of this chapter, you should be able to:

1. describe the chemical form of an acidic or basic ion exchange resin;
2. state that an ion exchange resin is a macromolecular substance which exchanges ionic units with ions in the surrounding solution;
3. explain the terms selectivity coefficient and distribution coefficient for an ion exchange material;
4. state that the ability of a resin to exchange ions with those in a dilute solution increases as the charge on the solvated ions increases;
5. set up and use an ion exchange column;
6. define the terms bed volume and exchange capacity;
7. describe the process of regeneration of a resin;
8. describe laboratory and industrial applications of ion exchange resins;
9. define chromatography as a means of separating mixtures by the distribution of its components between a stationary and a mobile phase in partition and adsorption chromatography;
10. describe the partition methods of paper, column, thin layer and gas chromatography;
11. define the terms R_f, R_x and retention volume;
12. describe, in outline, methods for the detection of colourless materials in paper and thin layer chromatography, and of eluents in GLC;
13. describe the process of electrophoresis;
14. describe the construction and function of a Soxhlet extractor;
15. state that it is more efficient to remove a solute from solution by using two or more portions of an immiscible solvent than to use the same total volume in one bulk;

16. describe the technique of solvent extraction;
17. describe how acidic and basic solvents can be used to extract basic and acidic materials respectively;
18. describe the use of chelation to extract an ionic substance into a non-polar solvent;
19. describe the technique of multistage extractions;
20. draw and set up the distillation apparatus suitable for simple distillation, for the evaporation of a flammable solvent, and for distillation under reduced pressure;
21. describe the principle and process of fractional distillation;
22. define an azeotrope as a mixture of constant boiling temperature, that is, it cannot be separated into its components by distillation;
23. describe the process of steam distillation and its application to the separation of the 2- and 4-nitrophenols;
24. list applications of the various distillation procedures to industrial processes.

Experiments

4.1 Measurement of the rate of exchange of ions between an ion exchange resin and the solution

Weigh out accurately approximately 1 g of a strongly acidic resin (H^+ form) and place it in a conical flask. Add deionised water (50 cm^3), 0.1 M sodium chloride solution (5 cm^3) and methyl orange indicator. Immediately stir the mixture. Add 0.1 M sodium hydroxide solution (1 cm^3) and start a stop clock. Note the time taken for the resin to neutralise the alkali, but do not stop the clock, and add a further 1 cm^3 portion of alkali. Repeat the procedure until there is no further change. Plot a graph of the volume of alkali added against the time for its neutralisation.

4.2 Determination of the capacity of an ion exchange resin

Take approximately 25 cm^3 of a strongly acidic resin (H^+ form) and place it in a measuring cylinder under deionised water. Stir and allow it to settle. Tap it down. Note the volume of the resin. Transfer it to a column and run through a 5 per cent w/v sodium chloride solution at a rate of 1 BV/2 minutes. Wash with water (50 cm^3) at the same rate. Collect the total eluent in a flask. Titrate the eluent against 0.1 M sodium hydroxide solution using screened methyl orange. Calculate the number of moles H^+ ion per dm^3 of resin.

4.3 Separation of chloride and bromide on an anion exchange resin

Make up a column of a strongly basic resin in the chloride form (40 g). Wash the column with 5 per cent nitric acid (about 200 cm^3) until the eluate gives no further positive test for chloride on addition of silver nitrate. Wash the column with 0.3 M sodium nitrate (50 cm^3) to displace the acid.

Prepare a solution of sodium chloride (0.1 g) and potassium bromide (0.2 g) in water (2 cm^3) and transfer it to the column. Elute the chloride with 0.3 M sodium nitrate solution at about 1 cm^3 per minute. Titrate 10 cm^3 portions of the eluate with 0.05 M silver nitrate solution using potassium chromate as indicator. Run a blank on 10 cm^3 sodium nitrate solution. When the titration value drops to near zero, elute the bromide with 0.6 M sodium nitrate solution.

Determine the percentage recovery of the chloride and bromide.

4.4 Illustration of the demineralisation of a solution

Connect two ion exchange columns, the first containing a strongly acidic resin (H$^+$ form) and the second a strongly basic resin (OH$^-$ form) (approximately 25 cm^3 of each). Prepare a solution which is 5.0×10^{-3} mol dm^{-3} in each of copper(II) sulphate and potassium dichromate and pass it through the two columns consecutively. Describe the result.

4.5 Catalysis by an ion exchange resin

$$CH_3 COOCH_2 CH_3 + H_2 O \rightleftharpoons CH_3 COOH + CH_3 CH_2 OH$$

The progress of this reaction may be followed by titrating the amount of acid present. As ethanoic acid is liberated, a higher titration value is obtained at each stage. To substantiate the catalytic effect, a 'blank' should be run, i.e. the experiment is run as indicated below but without the presence of the resin.

In a stoppered flask place a strongly acidic resin (H$^+$ form, 4 g) and deionised water (60 cm^3). Add ethyl ethanoate (2 cm^3), stir, start a stop clock and immediately remove a sample (20 cm^3) of the mixture. Further samples should be removed at 10 minute intervals up to 1 hour. If possible, remove a final sample 24 hours later.

Add each aliquot to a flask containing 50 cm^3 water, stir and titrate against 0.1 M sodium hydroxide solution using phenolphthalein as indicator. Plot the titre against time.

4.6 Preparation of a weak acid from its salt

Citric acid (2-hydroxypropan-1,2,3-trioic acid) can be prepared from its sodium salt by passing the salt through a weakly acidic resin.

Prepare a column of 25 cm^3 of the resin (H$^+$ form) and a solution of sodium citrate (50 mg) in water (100 cm^3). Pass the solution through the column at 4 cm^3/minute, and rinse the column with water (50 cm^3). Collect the eluent and evaporate to crystallisation. Cool and filter off the crystals. Determine the melting point.

4.7 Preparation of a silica sol

Prepare a column of a strongly acidic resin (H$^+$ form; 25 cm^3) and a 6 per cent solution of sodium silicate. Pass 50 cm^3 of the solution through the resin at 1 BV/15 minutes. The eluent is a silica sol.

4.8 Determination of the concentration of a metal ion in solution by ion exchange

Prepare a column of a strongly acidic resin (H$^+$ form; 10 cm^3) and a solution of a nickel salt (0.2 M in Ni^{2+} ions). Transfer the solution to the resin at 2 BV/minute. Wash the resin with deionised water and titrate the eluent with 0.1 M sodium hydroxide solution. Determine the concentration of eluted $H^+(aq)$ ions and so of Ni^{2+} ions in the original solution.

4.9 Purification of crude sugar

The removal of colouring matter (anionic) and other minerals (including cations) from sugar requires a mixed bed resin. Prepare a column from a weakly acidic (20 cm^3) and a strongly basic (10 cm^3) resin. The weakly acidic resin is used to prevent inversion of the sugar which is catalysed by a strongly acidic resin.

Prepare a solution (200 cm^3) containing 20 g of each of white and brown sugars. Pass through the column at 4 cm^3/minute. The sugar may be recovered from the eluent to determine the loss due to impurities.

4.10 Separation of metallic ions by paper chromatography

Paper chromatograms may be run as described on page 95 using either descending or ascending techniques.

Prepare solutions which are 1 M in $Pb^{2+}(aq)$, $Ag^+(aq)$ and $Hg_2^{2+}(aq)$, and a fourth solution which is 1 M in all three ions. Apply a spot of each to the paper (e.g. see Fig. 4.7) and develop with distilled water. Dry and then spray the paper with 0.25 M potassium chromate solution. Note the result. Wash the paper with a water spray and then expose it to ammonia fumes. What happens? Record the R_f values of the ions.

4.11 Separation of the ink components by two-way chromatography

Using the procedure indicated in Fig. 4.8, place a spot of black ink at the origin. Elute in one direction with a 3 : 1 : 1 mixture of butan-1-ol, ethanol and 2 M ammonia solution. Remove the paper, dry it, turn it through 90° and elute it with a 15 : 3 : 2 mixture of water, ethanol and saturated ammonium sulphate solution.

4.12 Separation of the isomeric nitroanilines by column chromatography

Prepare a column of alumina (25 g) in petroleum ether (60–80 °C fraction). Add a solution (5 cm^3) of the 2- and 4-nitroanilines (0.05 M in each). Develop the column using toluene (methylbenzene). Collect the separated isomers and evaporate to crystallisation. Determine the melting points of the crystalline products and so identify the order of elution.

4.13 Separation of amino acids by thin layer chromatography

A number of commercial systems are available for the preparation of TLC plates. In the absence of such a system, the following procedure can produce plates fairly effectively with practice.

Prepare a mixture of Kieselguhr G (3 g), or a similar silica gel material, and water (6 cm^3). Apply the mixture to microscope slides on a level surface (about 1 cm^3 per slide) and smooth it across the slides using the edge of another slide. Alternatively, apply the slurry by means of a spray; rinse the spray bottle out immediately as the silica rapidly hardens. The adsorbed water is then removed by heating in an oven at $100\,^{\circ}C$ for 30 minutes.

Remove the plates from the oven and allow them to cool in air. Apply spots of concentrated glycine, tyrosine, glutamic acid and aspartic acid solutions to the base lines. Run the plates in 7 : 3 ethanol–33 per cent ammonia solution mixture. Dry the plates, spray them with a ninhydrin solution (200 mg in 100 cm^3 propanone) and heat in an oven at $100\,^{\circ}C$. Determine the R_f values.

4.14 Separation by gas chromatography

The following experiment illustrates the use of a simplified gas chromatogram and the use of a chemical detection technique. If possible, supplement this experiment by using a commercial instrument (read the instruction manual) to separate the components of a suitable mixture.

The column should consist of a long glass tube bent into a U-shape and is filled with 'Tide' powder. The powder contains adsorbed water which acts as the stationary phase. The column will need to be immersed into a warm water bath. A gas inlet and sample injection inlet are required. Nitrogen is a suitable carrier gas, and a mixture of 1-aminopropane and 1-aminobutane should be injected on to the column. The eluent gas is passed from the column into water (25 cm^3), fresh samples of gas being collected in the water after suitable time intervals (e.g. 2 minutes). The aqueous medium is titrated against 0.01 M hydrochloric acid. The titre is plotted against the time.

The experiment can be repeated using varying gas pressures and column temperatures in order to determine the most suitable conditions for the separation of these amines. The temperature of the column may be varied by altering the temperature of the water bath.

Ethanol and propanol can be separated and detected by titration against 0.02 M potassium permanganate solution at $60\,^{\circ}C$.

4.15 Electrophoretic separation of a mixture of dyes

Set up an electrophoretic cell (Fig. 4.13, or a commercial alternative) using an ethanoate buffer of pH 10. Soak the paper in the buffer and spot on 1 per cent solutions of tartrazine, rhodamine, fluorescein and a mixture of all three.

Apply a potential of about 350 V (d.c.) and allow the separation to proceed. Dry the paper and comment on the result.

4.16 Recovery and purification of iodine using a Soxhlet extractor

Solutions containing iodide ions (e.g. from titrations) should be cautiously treated with 2 M sodium hypochlorite solution to precipitate out the iodine. Filter this off and wash with water. Allow the residue to dry in air and transfer to a Soxhlet extractor thimble (approximately half full). The solvent is tetrachloromethane. Iodine will crystallise out from the saturated solution.

4.17 Comparison of single and multiple liquid–liquid extractions

Prepare a solution of a water soluble dye (e.g. crystal violet) in water (50 cm^3).

To one portion of the solution (25 cm^3) in a separating funnel add trichloromethane (30 cm^3). Shake the mixture vigorously for 2 minutes, opening the tap occasionally to release the gas pressure. Allow the mixture to separate out into two liquid phases. Run off the lower organic layer into a flask and the upper aqueous layer into a beaker.

Extract the other portion (25 cm^3) with three successive portions (10 cm^3) as before. Collect the three organic fractions in a second flask, and the aqueous layer into another beaker.

Compare the intensities of the colours of the pairs of organic and aqueous layers in order to assess the effectiveness of the extraction process.

4.18 Purification by distillation

Using the apparatus of Fig. 4.19, distil a sample of impure ethyl ethanoate (with a trace of azobenzene colouring) (20 cm^3). A steam bath must be used for heating (the solvent is flammable); remember to add a suitable anti-bumping material (p. 109). Collect the condensate only when the temperature is steady. Record this temperature and the atmospheric pressure. What proportion of the mixture was recovered as pure ester?

4.19 Fractional distillation of a mixture

Prepare a 1 : 1 mixture of methylbenzene (toluene) and tetrachloromethane. Using a distillation apparatus with a suitable column (Fig. 4.22), add anti-bumping granules and heat gently. Collect fractions coming over below 82 $^\circ$C and then in 10° steps up to 112 $^\circ$C. Measure the volume of each fraction and plot a curve of the volume against temperature.

4.20 Distillation under reduced pressure

This may be achieved using the apparatus in Fig. 4.26. A faster separation can be achieved if a rotary evaporator is available; by rotating the flask, a larger surface area is produced, so accelerating the evaporation.

Ethyl ethanoylethanoate, a useful synthetic substance, is unstable to decomposition when distilled at atmospheric pressure. It is, therefore, more efficiently purified by distillation at reduced pressure.

Distil the crude material in the apparatus illustrated (Fig. 4.26) at 73–75 $^\circ$C and 2 kPa.

4.21 Purification of aminobenzene (aniline) by steam distillation

Using the apparatus shown in Fig. 4.25, purify crude aniline (10 cm^3) by adding water (10 cm^3) and passing steam through the hot solution. Collect the aniline/water condensate, add solid sodium chloride to reduce the solubility of aniline in water, and extract the aniline with ethoxyethane. Dry the ethereal solution over sodium hydroxide pellets, and evaporate the solvent off as in experiment 4.18 (or in a rotary evaporator). Redistil the aniline. Compare the boiling points of the crude and purified aniline.

References

Selected Experiments in Physical Science; D. H. Marrow (Longman, 1974).
An Introduction to Chemical Techniques; P. Tooley (J. Murray, 1975).
Ion Exchange Resins (BDH, 5th edn., 1977).
Ion Exchange: a Special Study (Nuffield Advanced Science; Longman, 1970).
Ion Exchange Experiments (Permutit Co. Ltd., 1968).
Analytical Applications of Ion Exchangers; J. Inczedy (Pergamon, 1966).
Ion Exchange; R. W. Thomas (Pergamon, 1970).
An Introduction to Ion Exchange; R. Paterson (Heyden/Sadtler, 1970).
An Introduction to Chromatography; D. Abbott and R. S. Andrews (Longman, 2nd edn., 1970).
Gas–liquid chromatography; D. R. Browning, *J. R. Inst. Chem.,* 1964, 376–81.
Gas Chromatography; R. S. Lowrie (Pergamon, 1969).
Construction of a Simple Gas–Liquid Chromatography Apparatus (ICI).
Chromatography: a Chemical Detective; D. R. Browning (Harrap, 1975).
Gel permeation chromatography; J. H. J. Peet, *Sch. Sci. Rev.,* 1971, **53**, 365–8.
Gel Filtration in Theory and Practice (Sephadex, 1971).
Gel chromatography; W. Heitz, *Angew. Chem. Inter. Edit.,* 1970, **9**, 689–702.
Modified gas–solid chromatography; J. J. Thompson and C. J. Heasman, *Sch. Sci. Rev.,* 1969, **50**, 536–48.
Dry column chromatography; B. Loev and M. M. Goodman, *Chem. Ind.,* 1967, 2026–32.

Films

An introduction to ion exchange (Permutit Co. Ltd.).
Chromatography (ICI).
Gas chromatography (Perkin Elmer).
Distillation (ICI).
Distillation (Shell).
Distillation of oil, parts I and II (BP).

Questions

1. Tap water can be softened by the use of a mixed bed ion exchanger. Explain the terms used in this statement and describe the process involved.

2. An analyst suspects that the peas in a tin of an imported product contain an illegal colouring material. Suggest how chromatography may be used to determine the validity of his suspicion.

3. The esterification product (0.62 g) of the reaction between ethanol and ethanoic acid was separated and detected by gas–liquid chromatography using octane (0.18 g) as an internal standard:

Substance	Retention distance/mm	Peak area/mm^2
Ethanol	78	260
Octane	86	310
Ethyl ethanoate	97	340
Ethanoic acid	108	230

Assuming that the detector responses to the standard and to ester are the same, and that this response is proportional to the number of moles, calculate the yield of ester in this reaction.

4. The R_f values of three substances A, B, C of a mixture in solvents 1, 2 and 3 are:

	1	2	3
A	0.51	0.49	0.82
B	0.76	0.73	0.50
C	0.75	0.57	0.51

For an adequate separation there must be a difference in the R_f values of at least 0.05. Which solvent is the most suitable? Sketch the result to scale.

5. Describe the process by which a concentrated solution of the grass pigments may be obtained for chromatography using a Soxhlet extractor.

6. Describe how distillation procedures are used in the oil industry.

Chapter 5

Analytical chemistry

Background reading

The background to this chapter is covered in *Fundamentals of Chemistry*, Chapters 4 and 12.

Titrimetric analysis

Titrimetric analysis is a quantitative determination of the components of a mixture by reacting measured volumes of two solutions together so that the reaction is just complete. For example, a specific volume of sodium hydroxide is taken (25.00 cm^3) and hydrochloric acid is added until the base has been neutralised; the volume used is noted.

Definitions

The solution of fixed volume is placed in a titration flask and is known as the **titrand**; the solution added from the burette is called the **titrant.** The **stoichiometric point** is the point of completion of the reaction according to the stoichiometry of the relevant equation. The progress of a titration is followed by an **indicator**; the indicator is used to determine the stoichiometric point, usually by a change in colour. The **end point** of a titration is the point at which the indicator shows the reaction to be complete. The **titration error** is the difference between the stoichiometric and practical end points.

It is essential in a titration that at least one solution is of exactly known concentration. This is the volumetric **standard**. This standard may be of a concentration determined from another titration, but there must be a substance in the analytical procedure which is a **primary standard**. A primary standard is a substance which can be converted into a standard solution by weighing the material out accurately on an analytical balance (to the nearest milligram or less) and dissolving it in water to obtain a solution of exactly known volume. To qualify as a primary standard, the substance must be pure; it must be of an exactly known composition (for example it must not effloresce or deliquesce); it must not react with the solvent, with air or with other reagents present in the solution. Typical primary standards are anhydrous sodium carbonate (an alkali), sodium tetrametaborate (an alkali), sulphamic acid (an acid), ethan-1,2-dioic acid ('oxalic acid', an acid or reducing agent), ammonium iron(II) sulphate (a reducing agent), and potassium iodate(VII) (an oxidising agent).

Sodium hydroxide is not suitable as a primary standard because of its hygroscopic properties and because it reacts with atmospheric carbon dioxide. Hydrated sodium carbonate loses water of crystallisation by efflorescence. Potassium permanganate reacts with bacteria in water and is reduced to manganese(IV) oxide. When a suitable primary standard has been chosen, it can be used to standardise other solutions; these, in turn, can be used as secondary standards.

Use of apparatus

1. Balances

A wide variety of chemical balances is available each with its own operating procedures. It is necessary, therefore, to leave the instructions for their use to an instructor on-the-spot. However, the principles may be described. It is important that an analytical balance be kept clean. Spillages should not occur. Material should not be added to, or removed from, the balance while the lever is free to move; it must be arrested first.

One of two methods may be used to measure accurately the mass of material used. (1) The empty weighing vessel is weighed accurately. The correct amount of material is added to the vessel (within the limits prescribed), and the apparatus is reweighed. The solid is then transferred to the beaker (or other suitable container), and the residual material is washed out of the weighing bottle and into the beaker. (2) Alternatively, the weighing bottle and approximate amount of solute are weighed out roughly. The container and solid are then weighed accurately. After the solid has been transferred to a clean beaker, the container is reweighed to determine exactly how much solute has been supplied.

Always carry out the weighing to the exclusion of draughts; close the front of the balance case and its side windows. Weigh the materials at the temperature of the balance, otherwise air currents are set up and introduce errors. Place objects in the centre of the pan. Remember that dirty hands, tongs, etc. introduce errors. Keep the balance clean and tidy.

2. Graduated flasks

Never transfer the solid directly to a graduated flask. If the solid is slow to dissolve, this flask cannot be warmed without destroying its reliability.

Dissolve the solute in a small amount of the solvent. For example, if a 250 cm^3 flask is to be used, 25–50 cm^3 solvent is adequate. Dissolution may be facilitated by stirring (take care to avoid splashing) or, *if* really necessary, by warming. The solution *at room temperature* is then transferred into the graduated flask through a clean filter funnel. The beaker, glass rod and funnel must be thoroughly washed with the solvent, the washings being collected in the flask. Any spillage invalidates the result and a fresh sample must be obtained.

Top up the solution with solvent to *near* the graduation mark. Add the final drops of solvent by means of a dropping pipette until the bottom of the meniscus just touches the graduation mark. Observe this at eye-level to avoid errors due to parallax. If you overshoot the mark – start again!

Stopper the flask, invert it several times (hold the stopper in!) so as to produce a homogeneous solution. The level of solution may apparently drop due to the sides of the flask being wetted; do *not* top it up.

3. Pipettes

A pipette is filled by drawing solution up by means of a pipette filler. Mouth suction must not be used because of the risk of poisoning by the chemicals and of contamination of the solution by saliva. Rinse out the pipette with the solution. Place the narrow tip of the pipette well below the surface of the liquid at room temperature and draw up the liquid into the pipette beyond the graduation mark. The solution should be allowed to run down to the graduation mark, with the bottom of the meniscus touching the mark. Again, this must be observed at eye-level. If mouth suction has been used, the liquid level is controlled by a finger being held over the upper end and the pressure being released slightly.

The outside of the tip of the pipette is then wiped with a filter paper but do not wipe across the open end as this would absorb some solution. Carefully move the pipette to the titration flask and allow the solution to run out freely – do not blow it out. When the liquid has ceased flowing, touch the pipette tip on the side of the vessel and hold it in the flask for 10 seconds. Then remove the pipette. Do not blow out the drop of liquid remaining in the pipette; the graduation has allowed for this.

4. Burettes

Burettes are normally graduated in 0.1 cm^3 steps. Rinse the burette out thoroughly with the appropriate solvent and ensure that it runs freely and is not blocked by dirt or grease. You cannot clean it out in the middle of a titration.

Clamp the burette vertically (not at an angle) in a burette stand. Fill it to above the zero mark by means of a funnel. Do not fill it too fast or else the burette will overflow. Aside from the wastage of materials and possible contamination of other materials, extra solution may enter the titration flask from outside the burette. Run the burette until the meniscus is on or just below the zero mark. Do not waste time trying to get a reading of 0.00 cm^3 – one drop of liquid may be too large for such precision. Read the initial level of solution to the nearest 0.05 cm^3. Run the titration and take the final reading. These readings should be taken at eye-level; a white card held behind the meniscus facilitates the reading.

The burette tap should be operated by the left hand — two fingers behind the burette and the thumb in front. The right hand should be used to gently swirl the flask to ensure mixing of titrant and titrand. Do not turn the tap with the right hand — this may result in the tap being loosened and the solution seeping out around the tap.

5. Cleanliness

Accurate and reproducible results cannot be obtained unless the glassware is clean. This can be determined visually by rinsing the apparatus with water; the water drains from clean glass leaving an even film (not droplets) on the wall of the apparatus. The procedure for cleaning apparatus is to remove the main contamination by washing with detergent and rinsing with water (any grease on the tap should be removed before application of the detergent). Final traces of grease can be removed by treatment with chromic acid. This is a very corrosive solution and it must not come into contact with skin or clothing. When using it, rubber gloves and goggles must be worn.

Acid—base titrations

An acid yields hydrogen ions ($H^+(aq)$) in aqueous solution. A base reacts with these hydrogen ions. Consider the reaction between hydrochloric acid ($H^+(aq)$ and $Cl^-(aq)$) and sodium hydroxide solution ($Na^+(aq)$ and $OH^-(aq)$):

$$Na^+(aq) + OH^-(aq) + H^+(aq) + Cl^-(aq) \rightarrow Na^+(aq) + Cl^-(aq) + H_2O$$

The net change is

$$H^+(aq) + OH^-(aq) \rightarrow H_2O$$

Some salts are able to act as acids or bases. For example, ammonium salts are acidic, the ammonium ion yielding a proton to the solvent:

$$NH_4^+ + H_2O \rightleftharpoons NH_3 + H_3O^+$$

Carbonates abstract protons from the solvent and so act as bases:

$$CO_3^{2-} + H_2O \rightleftharpoons HCO_3^- + OH^-$$

A useful salt is sodium tetrametaborate, or 'borax' as it is also known, ($Na_2B_4O_7 \cdot 10H_2O$) since this is a primary standard with basic properties.

$$B_4O_7^{2-} + 7H_2O \rightleftharpoons 4H_3BO_3 + 2OH^-$$

A range of indicators is available for these titrations. Each indicator changes colour over a range of pH. Figure 5.1 shows the working ranges of some indicators. The **working range** is the range of pH over which the colour changes. The actual value of the pH at the end point depends on the strengths of the acid and base. Figure 5.2 gives the titration curves (volume—pH graphs) for some typical acid—base mixtures. Curve 1 shows that at the stoichiometric point in the reaction between a strong acid and a strong base, the pH changes from 4 to 11 within a couple of drops of titrant. Nearly all the indicators in Fig. 5.1 change colour in this range and can be used to follow the titration.

128

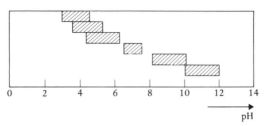

Fig. 5.1 Working ranges of some common indicators

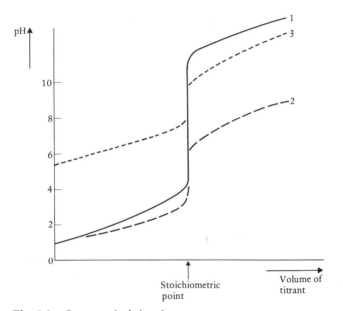

Fig. 5.2 Some typical titration curves

Strong acid–weak base mixtures give the curve 2. An indicator for this titration must change colour in a narrower range (pH 3.5–6.0). Suitable indicators for this titration would be methyl orange and bromocresol green. Curve 3 represents the change for a weak acid–strong base titration; phenolphthalein is a suitable indicator in this case.

The deviations of lines 2 and 3 from those of curve 1 can be related to the acid–base strengths. For example, in case 2, if the weak base is BOH, then the change is

$$H^+ + BOH \rightarrow B^+ + H_2O$$

Since the base is weak, an equilibrium is established, causing the cations to regenerate some hydrogen ions:

$$B^+ + H_2O \rightarrow BOH + H^+$$

The pH is, therefore, lower than demonstrated by the strong base situation.

When sodium carbonate is reacted with hydrochloric acid, two reactions occur:

$$Na_2CO_3 + HCl \rightarrow NaCl + NaHCO_3$$

and

$$Na_2CO_3 + 2HCl \rightarrow 2NaCl + CO_2 + H_2O$$

The first reaction is complete at a high pH and so can be followed by phenolphthalein. The second reaction has an end point at a low pH and is indicated by methyl orange.

Redox titrations

Redox, that is, reduction–oxidation titrations can be conducted using reagents which are coloured in the oxidised or reduced form but of a significantly different colour (or even colourless) in the other state. For example, potassium permanganate is purple in its oxidised form, but, after reduction to the manganese(II) ion, it is colourless.

$$MnO_4^- + 8H^+ + 5e^- \rightarrow Mn^{2+} + 4H_2O$$

In some cases the colour change is not sharp enough to ensure accuracy. An example is in the reaction of iodine with thiosulphate:

$$I_2 + 2e^- \rightarrow 2I^-$$
$$2S_2O_3^{2-} \rightarrow S_4O_6^{2-} + 2e^-$$
$$\text{(tetrathionate)}$$

The iodine is brown in solution; the iodide is colourless. As the iodine is consumed, the colour fades to a pale yellow which is difficult to distinguish from the colourless iodide ion. If starch is added as an indicator, an intense blue colour is formed in the presence of even a trace of iodine. It is colourless in the absence of iodine.

In some titrations an intensely coloured dyestuff is added in order to detect the end point. The dyestuff changes colour on reduction. For example, methylene blue is blue in its oxidised form and colourless when reduced. The controlling value (corresponding to pH in acidimetry) is the reduction potential.

Precipitation titrations

In some reactions a precipitate is formed. It is theoretically possible to conduct the titration until, at the end point, no further precipitate occurs on addition of the titrant. In practice this is very inaccurate. A number of alternative methods are possible, and these are illustrated for silver nitrate titrations.

1. The formation of a coloured precipitate (Mohr's method)
When a chloride is titrated against silver nitrate using potassium chromate as indicator, a white precipitate forms on the addition of the titrant. Not until the end point, because of a careful control of the concentration of the

indicator, does a brick red precipitate of silver chromate form. The concentration of chromate ion is selected on the basis of the relative solubility products (p. 43) of silver chloride and silver nitrate.

Silver chloride will precipitate from solution if the product of the ionic concentrations exceeds the solubility product:

$[Ag^+]$ $[Cl^-] > 2.0 \times 10^{-10}$ mol^2 dm^{-6} at 298 K

In a typical titration, 0.1 M solutions of silver nitrate (titrant) and sodium chloride (25.00 cm^3) might be used. So, $[Cl^-] = 10^{-1}$ mol dm^{-3} before the addition of Ag^+ ions. Precipitation will begin when

$$[Ag^+] > \frac{2.0 \times 10^{-10}}{10^{-1}} \text{ mol dm}^{-3}$$

i.e. 2.0×10^{-11}

One drop (about 0.05 cm^3) of silver nitrate would provide 5.0×10^{-6} mol Ag^+, or $\frac{5.0 \times 10^{-6}}{25.05} \times 1\,000 = 2.0 \times 10^{-4}$ mol dm^{-3} (the total volume of solution is 25.05 cm^3). This exceeds the concentration required; precipitation occurs immediately.

At one drop past the end point, the concentration of excess Ag^+ ions is $5.0 \times 10^{-6} \times \frac{1\,000}{50.0} = 1.0 \times 10^{-4}$ mol dm^{-3}. The solubility product of silver chromate is given by

$[Ag^+]^2 \cdot [CrO_4^{2-}] = 3.0 \times 10^{-12}$

So, for precipitation to just occur,

$$[CrO_4^{2-}] = \frac{3.0 \times 10^{-12}}{(1.0 \times 10^{-4})^2}$$

$$= 3.0 \times 10^{-4}$$

The normal amount of the potassium chromate indicator used is 1 cm^3 of a 0.25 M solution. This contains 2.5×10^{-4} moles, which is within experimental error of the calculated figure.

2. The formation of a coloured solution (Vohlhard's method)
In this case a red solution of an iron(III)–thiocyanate complex is formed at the end point of a silver nitrate–ammonium thiocyanate titration using iron (III) alum as indicator. The amount of indicator used can be determined by comparison of the equilibrium constants as described for the Mohr method above.

3. The use of adsorption indicators (Fajan's method)
When chloride ions are determined by silver nitrate solution in the presence of fluorescein, a red colour forms on the silver chloride precipitate at the end point. Before the end point the excess chloride ions are adsorbed on to silver chloride precipitate; past the end point the silver ions are preferentially

adsorbed. When there is an excess of silver ions, a silver salt is formed on the surface of the silver chloride colloid: the red silver fluoresceinate.

Calculations

Titration calculations can be reduced to a few simple steps. It is essential to know which substance is acting as the standard. The steps required are: (1) the determination of the number of moles of the standard used; (2) the writing of a balanced chemical equation; (3) the calculation of the number of moles of the other compound; (4) the conversion of the number of moles of the other compound to a mass; and so (5) the determination of the quantity (concentration, purity, etc.) required. Some examples follow to illustrate the method of calculation.

Example 1

A solution of potassium permanganate (3.10 g dm^{-3}) was titrated against ethan-1,2-dioic acid ($H_2C_2O_4 \cdot 2H_2O$). 25.00 cm^3 of the latter solution were oxidised by 18.20 cm^3 of the permanganate solution to carbon dioxide. What is the concentration of the ethan-1,2-dioic acid in grams per dm^3?

The relative molar mass of $KMnO_4$ = 158.03

Molarity of the potassium permanganate solution = $3.10/158.03$ M

Number of moles of potassium permanganate used in the titration

$$= \frac{18.20}{1\,000} \times \frac{3.10}{158.03}$$

$$= 3.570 \times 10^{-4}$$

Equations for the reaction:

$$MnO_4^- + 8H^+ + 5e^- \rightarrow Mn^{2+} + 4H_2O$$

$$C_2O_4^{2-} \rightarrow 2CO_2 + 2e^-$$

So, 2 moles of permanganate react with 5 moles of ethan-1,2-dioate. Therefore, in the titration $3.570 \times 10^{-4} \times 5/2$ moles of ethan-1,2-dioate were consumed.

1 dm^3 of ethan-1,2-dioate solution therefore contains

$$\frac{1\,000}{25.00} \times 3.570 \times 10^{-4} \times \frac{5}{2} \text{ moles}$$

i.e. it is $0.035\,7$ M.

Since the relative molecular mass of the acid is 126.06, the concentration of the acid solution is $0.035\,7 \times 126.06$ g dm^{-3}, that is 4.50 g dm^{-3}.

Example 2

Potassium iodide (8.50 g) was dissolved in water to give a homogeneous solution (1 dm^3). 25.00 cm^3 of this solution was titrated with 0.05 M silver nitrate solution using an adsorption indicator. The end point was reached at 22.50 cm^3. Calculate the percentage purity of the potassium iodide.

The primary standard is the silver nitrate.

Number of moles of silver nitrate used in the titration

$$= \frac{22.50}{1\,000} \times 0.05$$

$$= 1.125 \times 10^{-3}$$

Equation of reaction:

$$Ag^+ + I^- \rightarrow AgI\ (s)$$

that is, they react in a 1 : 1 ratio.
So, the number of moles of KI must also be 1.125×10^{-3}.

Therefore, the molarity of the KI solution $= \dfrac{1\,000}{25.00} \times 1.125 \times 10^{-3}\ M$

$$= 0.045\ M$$

Since the relative molecular mass of KI is 166.00,
the concentration is $0.045 \times 166\ g\ dm^{-3}$ $= 7.47\ g\ dm^{-3}$
Hence the purity of the potassium iodide

$$= \frac{\text{concentration of the pure salt}}{\text{concentration of the impure salt}} \times 100\ \text{per cent}$$

$$= \frac{7.47}{8.50} \times 100\ \text{per cent}$$

$$= 87.9\ \text{per cent}$$

Example 3

Crystalline sodium carbonate ($Na_2CO_3 \cdot xH_2O$) (4.29 g) was made up to
250 cm³ of solution, and 25.00 cm³ of this solution required 15.00 cm³ of
0.20 M hydrochloric acid for neutralisation. Calculate the value of x, the
number of moles of water of crystallisation per mole of sodium carbonate.

The standard is the acid.

Number of moles of acid used in the titration $= \dfrac{15.00}{1\,000} \times 0.20$

$$= 3.000 \times 10^{-3}$$

Equation:

$$CO_3^{2-} + 2H^+ \rightarrow CO_2 + H_2O$$

that is, 1 mole carbonate reacts with 2 moles hydrogen ions.

So, number of moles of carbonate $= \dfrac{1}{2} \times 3.0 \times 10^{-3}$

$$= 1.50 \times 10^{-3}$$

Therefore, number of moles of carbonate per 250 cm³ $= 1.500 \times 10^{-3} \times 10$

$$= 1.500 \times 10^{-2}$$

The relative atomic mass of $Na_2CO_3 \cdot xH_2O$ $= (105.97 + 18.01x)$

The mass of carbonate per 250 cm³ $= 1.50 \times 10^{-2} \times (105.97 + 18.01x)$

So, $1.50 \times 10^{-2}\ (105.97 + 18.01x) = 4.29$

or, $x = 10.00$

Therefore, the number of moles of water of crystallisation per mole of sodium carbonate is 10.0.

Electrometric titrations

Conductimetric titrations

The variation in conductance with the ionic concentration provides a means of titrimetric analysis. Consider the neutralisation of sodium hydroxide solution by hydrochloric acid. If the cell contains sodium hydroxide solution only, the conductance is relatively high. As acid is added,

$$Na^+(aq) + OH^-(aq) + H^+(aq) + Cl^-(aq) \rightarrow Na^+(aq) + Cl^-(aq) + H_2O$$

the conductivity drops due to the removal of hydroxide ions as covalent water molecules and their replacement with chloride ions which have a lower conductance (199.1 and 76.3 ohm^{-1} m^{-2} mol^{-1} respectively). At the end point the conductance reaches a minimum and increases thereafter due to the build-up of excess hydrogen ions (Fig. 5.3).

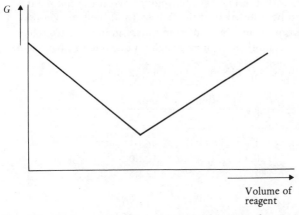

Fig. 5.3 Progress of a conductimetric titration of a strong acid with a strong base

If a weak base is used instead of sodium hydroxide, the conductance is less initially, and the addition of acid increases the conductivity up to the end point as cations and anions replace undissociated molecules. After the end point, the conductivity rises more rapidly due to excess ions of the

$$NH_3(aq) + H^+(aq) + Cl^-(aq) \rightarrow NH_4^+(aq) + Cl^-(aq)$$

strong acid (Fig. 5.4).

Table 5.1 gives the molar conductances of some common reagents. From

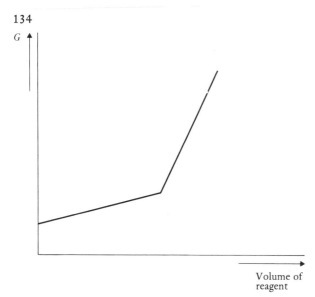

Fig. 5.4 Progress of a conductimetric titration of a strong acid and a weak base

these values it is obvious that in Fig. 5.3 the conductance at the end point (Na^+Cl^- solution) must be below that of the free acid and the base. On the other hand, neutralisation of ammonia solution causes an increase from 9.66 to 122.5 units at 291 K, and then to an even higher value due to excess acid.

Table 5.1 Molar conductance of some common reagents

Electrolyte (0.01 M in water)	$\Lambda_o \times 10^4$/ohms^{-1} m^2 mol^{-1}	
	298 K	291 K
HCl	411.8	368.1
NaCl	118.45	102.0
NH_4Cl	141.21	122.5
NaOH	237.9	
CH_3COOH		14.50
NH_4OH		9.66

The conductance is followed by using a conductance bridge (which is calibrated in ohms^{-1} (Ω^{-1}, Siemens)) or by a Wheatstone bridge. The latter requires an a.c. power supply and a suitable detector for the balance point (see *Fundamentals of Chemistry*, p. 140). The various commercial conductance bridges operate differently, and the manufacturers' instructions must be followed in order to determine the correct operation. The conductance cells should be handled with care in view of their delicate nature, and stored in conductivity water (i.e. deionised water). The conductance is dependent on the temperature, and so the temperature of the solutions should be regulated by a thermostat bath.

Calculation

A typical experiment gave the following results in the titration of 0.100 M ethanoic acid against sodium hydroxide solution (25.00 cm^3). Calculate the molarity of the solution.

Volume of base/cm^3	Conductance/ 10^4 μohm^{-1}	Volume of base/cm^3	Conductance/ 10^4 μohm^{-1}
0.0	1.00	22.0	0.25
2.0	0.92	24.0	0.25
4.0	0.84	26.0	0.24
6.0	0.77	28.0	0.23
8.0	0.69	30.0	0.23
10.0	0.61	32.0	0.22
12.0	0.52	34.0	0.21
14.0	0.45	36.0	0.20
16.0	0.37	38.0	0.20
18.0	0.29	40.0	0.19
20.0	0.26		

Figure 5.5 gives a plot of the results listed. The point of intersection corres-

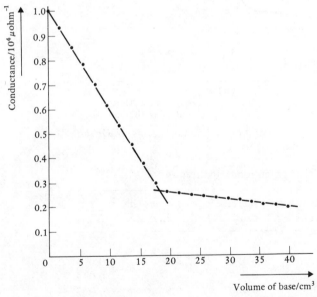

Fig. 5.5 Conductimetric titration of ethanoic acid with sodium hydroxide

ponds to 18.8 cm^3 of the acid. So, the number of moles of acid used are $18.8 \times 10^{-3} \times 0.100 = 1.88 \times 10^{-3}$. Since 1 mole of acid reacts with 1 mole of base,

$$CH_3COOH + NaOH \rightarrow CH_3COONa + H_2O$$

136

25.00 cm^3 base also contains 1.88×10^{-3} moles. Therefore, the molarity of the sodium hydroxide solution is $1.88 \times 10^{-3} \times 1\,000/25.0 = 7.52 \times 10^{-2}$.

Potentiometric titrations

A variation in cell e.m.f. with a change in concentration of the electrolytes provides another means of titration. Consider the titration of sodium hydroxide solution with hydrochloric acid. Figure 5.6 illustrates the change in e.m.f. with the volume of added base. The end point is the mid-point of the almost vertical line (that is, at the point of inflection).

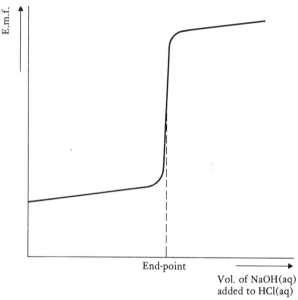

Fig. 5.6 Potentiometric titration of HCl (aq) against NaOH(aq)

Table 5.2 Results of potentiometric titration

Vol. of KmnO$_4$/cm^3	E.M.F./ volts	Increase in e.m.f., $\triangle E$	Increase in vol., $\triangle V$	$\dfrac{\triangle E}{\triangle V}$
1.0	0.726	—	—	—
5.0	0.782	0.056	4.0	0.014
9.0	0.839	0.057	4.0	0.014
9.9	0.899	0.060	0.9	0.067
9.95	1.000	0.101	0.05	2.02
10.0	1.468	0.468	0.05	9.36
10.1	1.485	0.017	0.1	0.17
11.0	1.497	0.012	0.9	0.013
12.0	1.501	0.004	1.0	0.004

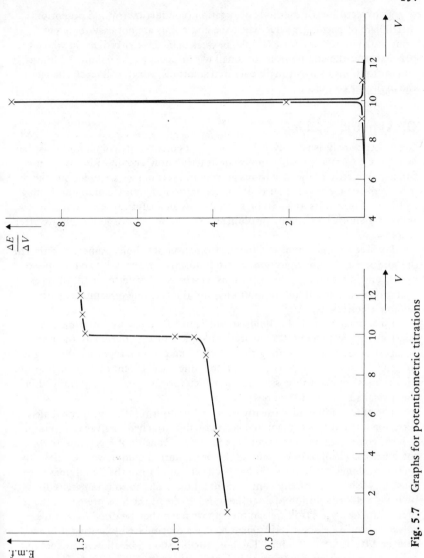

Fig. 5.7 Graphs for potentiometric titrations

It is often difficult to determine exactly the position of the end point on such a graph. It can be found with a high degree of confidence using the 'first differential curve'. This is a mathematical concept which indicates that $\Delta E/\Delta V$ is a maximum at a point of inflection. $\Delta E/\Delta V$ is the gradient of a graph of E (e.m.f.) against V (volume). The method of determination is indicated in Table 5.2 for a titration between iron(II) ions and permanganate ions. Figure 5.7 gives the e.m.f.–volume and first differential curves for this experiment.

The electrometric methods of titration (conductimetric and potentiometric) are of particular value when the end points are not easy to detect using visual indicators, for example, in weak acid–base titrations, in which the pH change at the end point is too small with respect to the volume additions, and in titrations using strongly coloured solutions, which will mask changes in the indicator colours.

Gravimetric analysis

Gravimetric analysis is analysis by mass. An element or its compound must be in as pure a form as possible and of an exactly known composition to be useful in gravimetric analysis. Various gravimetric techniques are used, but the most versatile method is that of the precipitation of a pure compound from a solution of one of its ions. To be successful as an analytical technique, the precipitate must be virtually insoluble so that none remains in the filtrate after filtration.

Possible complications include the formation of colloids, supersaturation of the solution and coprecipitation. These problems can generally be minimised by: (i) precipitation from hot solutions which are allowed to stand; (ii) precipitation from stirred, dilute solutions; and (iii) washing precipitates with suitable electrolytes.

Fine precipitates and colloids are difficult to remove by filtration and so relatively coarse precipitates are preferable. Coagulation occurs by the process of digestion, that is, by heating the solution and precipitate, or by allowing it to stand for several hours. Digestion also reduces the extent of coprecipitation. This term is applied to the situation in which foreign ions become trapped in the crystal structure of the precipitate.

The pure compound is usually isolated by filtration. The most convenient filtration apparatus is the sintered glass crucible attached to a vacuum pump. It is, of course, essential to ensure the complete transfer of a precipitate to the crucible. The beaker containing the precipitate should be washed clean by several washings. Once the solid has been separated from the liquid phase, it must be washed free of any contaminants. Losses due to solution are reduced by using a dilute solution of a salt containing one of the ions present in the precipitate (see p. 44). The clean precipitate must then be dried to remove solvent. In many cases it is sufficient to dry a precipitate by heating it in an oven at $110\,^{\circ}$C for 1–2 hours. The hot crucible should be allowed to cool in a desiccator to room temperature before weighing (p. 125). The desiccant is normally silica gel or anhydrous calcium chloride. Alternatives include concentrated sulphuric acid and phosphorus(V) oxide. The choice of desiccant is often dependent on the nature of the precipitate; no reaction must occur between them.

When the mass of precipitate has been determined, the solid should be reheated, cooled and weighed. The cycle should be repeated until consecutive masses are within 0.2 mg of each other.

If the final form of the substance is obtained by heating the precipitate at a high temperature (e.g. an oxide from a hydroxide), a ceramic crucible is used. Nickel or platinum crucibles are used for high temperature reactions involving bases and acids. Ceramic and metal crucibles are supplied with lids to prevent loss of important solid as smoke, and yet to allow the intake of oxygen, if required, for combustion. The precipitate is collected in a filter paper and the paper plus precipitate is transferred to the crucible. The crucible is heated gently to drive off moisture, and then more strongly to burn off the paper and finally to convert the solid to the required form. Ashless filter papers are used, since they burn away leaving no solid residue.

An alternative to sintered glass crucibles is the Gooch crucible — it is a ceramic crucible with a base containing a number of holes. The base is covered with an asbestos pad for filtration purposes. It has the advantage, over the sintered glass crucible, of being usable at high temperatures; its advantage over the other crucibles is that it can be used in the filtration process.

Calculation

The process of calculation may be illustrated by the case of an iron compound (1.000 g) which was dissolved in water, treated with base to precipitate the hydroxide which was then heated to constant mass, yielding iron(III) oxide (0.156 5 g).

1 mole Fe_2O_3 contains 2 moles Fe atoms.

So, 159.691 g oxide contains 111.694 g iron.

Therefore, 0.156 5 g oxide contains $\dfrac{111.694}{159.691} \times 0.156\,5$ g iron

i.e. 0.109 4 g iron

This iron all originated from the compound, so the percentage of iron in the compound is $\dfrac{0.109\,4}{1.000} \times 100 = 10.94$ per cent

Spectrophotometric and related methods

When atoms are irradiated with electromagnetic radiation (*Fundamentals of Chemistry*, p. 3), energy is absorbed causing electrons to be promoted to higher energy levels. Molecules have a larger range of possibilities for the absorption of irradiated energy (Table 5.3). Aside from the electronic transi-

Table 5.3 Effects of absorbed electromagnetic radiation

Energy/ kJ mol^{-1}	10^8	10^6	10^4	10^2	1	10^{-2}	10^{-4}
Description	Gamma	X-radiation	U.V. visible	I.R.		Microwaves	Radiowaves
Absorption process	Nuclear transitions	Inner electron transitions	Outer electron transitions	Vibrational changes		Rotational; electron spin orientations	Nuclear spin orientations

tions, energy absorption can cause molecular vibrations or rotations. Energy absorption can also affect the nuclear energy and rotational energies of the nuclei and electrons.

The various spectral regions can be used for analytical purposes. The basic instrumental layout in these instruments is shown in Fig. 5.8. In the

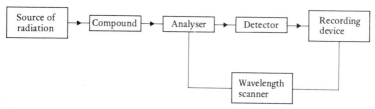

Fig. 5.8   Basic layout of a spectrophotometer

case of visible light spectroscopy, a tungsten lamp is used as the radiation source; a hydrogen discharge lamp emits ultraviolet radiation and an electrically heated filament generates infrared radiation. The radiation is passed through a sample of the compound. In order to determine which absorptions have occurred, the radiation is passed into an analyser. This spreads the wavelengths out into an orderly band; for example, the rainbow-like spectrum is produced when white light is passed through a glass prism. The radiation is detected and converted to an electrical signal which is recorded on a chart recorder. By means of a scanner, the pen of the recorder is regulated to respond to the intensity of the various wavelengths of the radiation distinguished by the analyser.

Colorimetry

Many compounds, both inorganic and organic, are coloured (see *Fundamentals of Chemistry*, pp. 133 and 210). In solution, the intensity of colour is related to the concentration of the solution, as shown in the equation

$$\log \frac{I_0}{I_t} = \epsilon \cdot c \cdot l$$

where I_0 and I_t are the intensities of the incident and transmitted lights respectively; ϵ is a constant (**absorptivity**); c is the concentration and l, the length of the medium (that is, the distance the light travels through the solution). This is known as the **Beer–Lambert law**. The intensity of the light is measured by a photoelectric cell, though versions of the colorimeter are available in which the light intensity is measured visually by comparison with standards. The photoelectric cell is usually employed with light covering a narrow range of wavelengths. If a white light source is used, a filter is incorporated to eliminate unwanted light. The filters must be constructed of glass, gelatin or other transparent materials impregnated with dyes. The filter is selected by taking readings using each of the spectrum filters in turn. Generally, the filter which gives the maximum absorption for the substance for a chosen concentration is the best one to use. The colour of the filter is the complementary colour to that of the solution (Table 5.4).

Table 5.4 Complementary colours for filters

Solution colour	Filter colour
Violet	Yellow − green
Blue	Yellow
Green − blue	Orange
Blue − green	Red
Green	Purple
Yellow − green	Violet
Yellow	Blue
Orange	Green − blue
Red	Blue − green

The term $\log (I_0/I_t)$ is often called the **optical density**, or **absorbance**, A. An alternative term is that of **transmittance**, T, where $T = I_t/I_0$. So, $A = \log (1/T)$.

Many substances have suitable colours to permit their direct estimation, but it is often necessary to add a reagent which complexes with the material under investigation to give a coloured solution. For example, iron(III) ions produce an intense red colour with thiocyanate ions; manganese(II) ions are most easily examined after oxidation to permanganate ions.

Determination

The normal procedure is to produce a standard curve relating the intensity of the transmitted light (or optical density) to the concentration of a series of standard solutions. These are plotted as a graph of the absorbance (A) or of the log (transmittance) against the concentration. The graph is known as the **calibration curve** for the instrument. Light of the same intensity and wavelength is passed through a solution of unknown concentration. The absorbance or transmittance is measured and, by use of the standard graph, is converted to a concentration term.

For example, 1.00 g of an ore containing chromium is oxidised in 100 cm^3 solution to give the coloured dichromate ion. The solution is found to have an optical density of 0.55 when measured in a colorimeter. A series of standard solutions of chromium give the tabulated results in the same colorimeter. What is the concentration of chromium in the ore?

Concentration/g dm^{-3}	Optical density
0.04	0.20
0.06	0.30
0.08	0.40
0.12	0.60
0.16	0.80

The standard values are plotted to give the results shown in Fig. 5.9. The graph is linear because, in accordance with the Beer−Lambert law, the optical density is proportional to the concentration of the absorbing species.

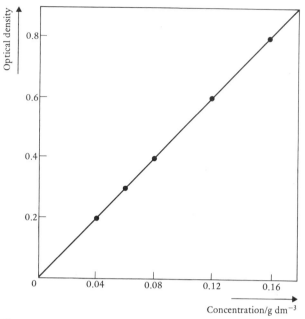

Fig. 5.9 Calibration curve for the Beer–Lambert law

An optical density of 0.55 corresponds to a concentration of 0.11 g dm⁻³
The concentration of the ore is 10.0 g dm⁻³, so the percentage of chromium

in the ore is $\dfrac{0.11}{10.0} \times 100 = 1.1$ per cent.

Sources of inaccuracy

To be a satisfactory method of analysis, the colour used must be specific
to the species investigated. If two substances give similar colours (that is,
transmitted light of similar wavelengths), it is not possible to distinguish
between them colorimetrically. The technique assumes the validity of the
Beer–Lambert law. This is inapplicable if there is a change in its chemical
nature as the concentration changes (due, for example, to dissociation or
association). The colour must be stable over the period of measurement.
Some compounds are oxidised by air on standing and so the colour will
change. The solution must be free of any precipitate (or colloid) which will
affect the transmittance of light due to scattering.

Ultraviolet–visible spectrophotometers

Many types of instrument are available now for the measurement of
absorption in the ultraviolet and visible regions of the spectrum. These
regions are conveniently examined together. A hydrogen discharge lamp is
used as a source of ultraviolet radiation; visible light is produced by a
tungsten lamp. Since glass is not transparent to ultraviolet radiation, silica

(which is transparent up to about 1.8×10^{-7} m) optical parts are used in preference to glass. Diffraction gratings may be used instead of prisms. Normally a double beam instrument is employed – one beam passing through a reference cell (say, the solvent) and the other through the sample. The recorder plots the difference in absorption between the two cells. The spectrophotometer (in contrast to the colorimeter) records the absorbance over a range of wavelengths. Single wavelengths may be employed.

The cells should be handled as little as possible (and then only at the ground glass faces). Any finger prints etc. on the transparent faces will affect the intensity of the transmitted light.

Infrared spectrometry

The infrared spectra are widely used as a means of identifying the presence of particular bonds. Each chemical bond absorbs at a specific energy, the precise values being determined by the chemical environment. Figure 5.10 gives a typical spectrum, that of propanone (acetone).

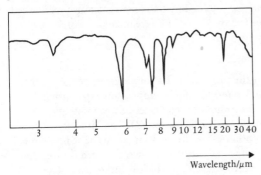

Wavelength/µm

Fig. 5.10 Infrared spectrum of propanone

As a general guide, Table 5.5 shows how the energy of absorption is related to bond order. Tables can be found in analytical works which enable a more specific indication of the bond. Infrared spectroscopy can also be used for quantitative analysis since the intensity of absorption is proportional to the concentration of the absorbing species.

Table 5.5 Some typical infrared absorption values

Wavelength/µm	2.7–3.2	4.0–5.0	5.5–6.5	7.5–12.0
Bond type	X–H	C≡X	C=X	C–X
Example	O–H	C≡N	C=O	C–Cl
Approximate wavelength/µm	2.75	4.55	6.15	14.80

The infrared spectrophotometer uses an electrically heated filament (**Nernst filament**) based on thorium oxide. The radiation from this source is dispersed using a rock salt prism (which transmits to 15 µm), glass being opaque to this region of the spectrum. Other salts may be used if greater ranges or higher resolutions are required. Modern instruments use diffraction gratings. The intensity of the transmitted radiation is measured using a thermocouple.

Infrared spectra may be run in any physical state. Most commonly solids are examined as a dispersion in 'nujol', a non-volatile liquid alkane. A small amount of the solid is ground with nujol and a thin layer of the paste is smeared on to a rock salt disc. A second disc is added to spread the paste out evenly. The spectrum contains absorptions due to the nujol. These can be used as 'markers' for checking the correctness of the mounting of the chart, but they must be allowed for in the interpretation of the spectra. Alternatively solids may be converted into transparent discs by compressing a mixture of the solid and potassium bromide. This technique is more difficult but avoids the presence of absorption peaks due to the support.

Liquids can be used in cells with rock salt windows. However, aqueous solutions must not be put into these cells – sodium chloride dissolves in water! Calcium fluoride cells may be used instead.

Rock salt plates should not be touched by hand, and must not be washed with an aqueous solvent. After use they should be cleaned and stored in a vacuum desiccator.

Flame photometry

Flame photometry is based on the production of colour by atoms, or their ions, in a flame. The atoms are heated to a high temperature, so causing their electronic excitation. In returning to their states of lower energy, they emit radiation (see *Fundamentals of Chemistry*, p. 13). The intensity of the emitted light is measured by a photoelectric cell. Extraneous light is excluded by the use of filters. The flame normally used is a town gas/air mixture, the relative pressures being regulated to give a flame of low luminosity. The sample is passed into the flame by means of an atomiser. The photoelectric cell response is recorded by a sensitive galvanometer (Fig. 5.11).

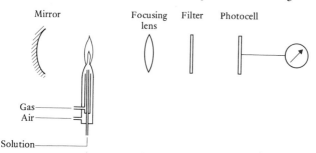

Fig. 5.11 Diagrammatic representation of a flame photometer

Calculation

A cement (1.000 g) was analysed for sodium and potassium content by solution in acid and dilution with water (to 100 cm^3). Standard solutions of sodium and potassium in the same medium were prepared and the various solutions were examined in a flame photometer. The results are given in Table 5.6.

Table 5.6 Flame photometer readings

| Concentration/μg cm^{-3} | Galvanometer readings | |
	Na	K
0	9	0
25	34	26
50	60	53
70	80	73
Cement	38	58

The calibration curves are shown in Fig. 5.12. From these curves it can be seen that the cement contains 29 μg Na and 55 μg K per cm^3 solution. Percentage values can be derived. 29 μg Na per cm^3 indicates that 1.000 g cement contained 29 × 10^{-6} × 100 g sodium (1.000 g dissolved in 100 cm^3). So, the

percentage of sodium in the cement is $\dfrac{29 \times 10^{-6} \times 100}{1.000} \times 100 = 0.29$ per

cent. Similarly, it contains 0.55 per cent potassium.

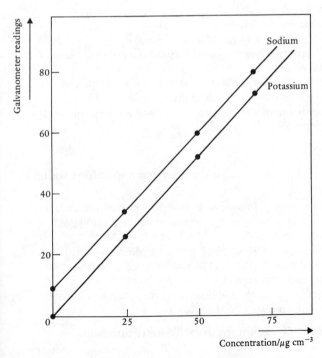

Fig. 5.12 Calibration curves for a flame photometer

Summary

At the conclusion of this chapter, you should be able to:

1. state what is meant by a titration;
2. define the terms used in titrimetry;
3. name some suitable primary standards;
4. describe why sodium hydroxide, hydrated sodium carbonate, and potassium permanganate are not suitable for use as primary standards;
5. use correctly such apparatus as a balance, graduated flask, pipette and burette;
6. prepare a solution of a primary standard of exactly known molarity;
7. carry out accurately a titration;
8. describe how the end point may be obtained for an acid−base, redox and precipitation titrations;
9. calculate the molarity of a solution from the titration results, and hence any other required quantity (e.g. concentration in g dm^{-3}, purity, percentage composition);
10. describe the principles and procedures of conductimetric and potentiometric titrations;
11. describe the principles and procedures in gravimetric analysis;
12. calculate the percentage composition of a substance from a set of gravimetric results;
13. state the Beer−Lambert law;
14. define the optical density and transmittance;
15. describe the colorimetric analysis of a substance;
16. calculate the percentage composition of a substance from the results of colorimetric analysis;
17. describe the principles and application of ultraviolet−visible spectrophotometers and infrared spectrophotometry;
18. describe the operation of a flame photometer and interpret the results of such an experiment.

Experiments

5.1 Standardisation of hydrochloric acid using anhydrous sodium carbonate

Prepare a standard solution of sodium carbonate by weighing out accurately between 1.3 and 1.4 g of anhydrous sodium carbonate (analytical grade reagent) and making it up to 250.0 cm^3 of solution (p. 125).

Titrate the carbonate (25.00 cm^3) against the hydrochloric acid using 2−3 drops of screened methyl orange as indicator (p. 127). Repeat the titration to obtain concordant results.

Repeat the titration using phenolphthalein as indicator. Comment on the results. Calculate the molarity of the acid.

5.2 Standardisation of sodium hydroxide solution against sulphamic acid

$$NH_2 SO_3 H + NaOH \rightarrow NH_2 SO_3 Na + H_2 O$$

Prepare a standard solution of sulphamic acid using between 2.4 and 2.5 g in 250 cm^3 solution. Titrate the base (25.00 cm^3) against the acid using phenolphthalein. Calculate the concentration of the base in g dm^{-3}.

5.3 Standardisation of a potassium permanganate solution

$$5Fe^{2+} + MnO_4^- + 8H^+ \rightarrow 5Fe^{3+} + Mn^{2+} + 4H_2O$$

Prepare a standard solution (250 cm^3) of ammonium iron(II) sulphate (approximately 9.8 g).

Pipette a portion of this solution (25.00 cm^3) into a titration flask, add dilute sulphuric acid (10 cm^3) and titrate against the permanganate solution to the first permanent pink colour. Determine the molarity of the permanganate.

5.4 Determination of the percentage of ethan-1, 2-dioate (oxalate)

Prepare a solution (250 cm^3) of the sodium ethan-1, 2-dioate (1.7 g, determined accurately). Transfer 25.00 cm^3 to a clean flask, add 2 M H_2SO_4 (10 cm^3), heat to 70 °C and titrate to the first permanent pink colour. The temperature must not fall below 60 °C during the titration. Calculate the percentage of ethan-1, 2-dioate ($C_2O_4^{2-}$) in the sodium oxalate.

$$5C_2O_4^{2-} + 2MnO_4^- + 8H^+ \rightarrow 10CO_2 + 2Mn^{2+} + 4H_2O$$

5.5 Determination of the percentage of copper in copper(II) sulphate

$$2Cu^{2+} + 4I^- \rightarrow 2CuI(s) + I_2$$

$$I_2 + 2S_2O_3^{2-} \rightarrow 2I^- + S_4O_6^{2-}$$

Weigh out accurately approximately 3.0 g copper(II) sulphate pentahydrate and make up to 250 cm^3 solution. Pipette 25.00 cm^3 into a titration flask, add potassium iodide (1 g) and titrate the liberated iodine against 0.100 M thiosulphate solution. As the brown colour of the iodine is discharged, a white precipitate of copper(I) iodide is apparent. When the solution has become pale yellow, add starch solution (1 cm^3) and complete the titration until a colourless solution is obtained (the precipitate remains).

Calculate the concentration of copper(II) ions, and so the percentage of copper in the salt.

5.6 Determination of the percentage of chlorine in 'common salt'

Weigh out accurately approximately 1.5 g sodium chloride and prepare a 250 cm^3 solution. To 25.00 cm^3 of this solution, add potassium chromate solution (1 cm^3) and run in 0.100 M silver nitrate solution until the first permanent red precipitate is formed.

Calculate the percentage of chlorine in the salt.

5.7 Conductimetric titrations

Using the pairs of solutions indicated, place 25.0 cm^3 of one solution in the cell and determine the conductance. Add 2.0 cm^3 of the other solution and redetermine the conductance. Measure the conductance after the addition of further 2.0 cm^3 portions up to a total of at least 40.0 cm^3.

Cell	Burette
1 M NaOH	1 M HCl
1 M NH$_3$	1 M HCl
1 M NaOH	1 M CH$_3$COOH
1 M NaOH	1 M Glycine
0.1 M NaCl	0.1 M AgNO$_3$

5.8 Potentiometric titration of iron(II) and permanganate

The apparatus consists of a beaker containing the titrand, a platinum electrode and a calomel electrode. The electrodes are connected to a potentiometer circuit (Fig. 5.13) in order to determine the e.m.f. of the cell.

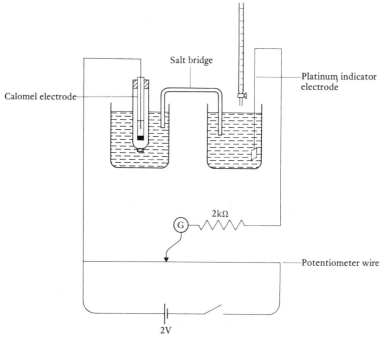

Fig. 5.13 Circuit for potentiometric titration

Place 0.100 M iron(II) sulphate solution (25.00 cm^3) in the cell, and acidify with 2 M sulphuric acid. The cell solution should be stirred continuously during the experiment. Determine the e.m.f. Add approximately 0.02 M potassium permanganate solution in 3.0 cm^3 steps up to 21.0 cm^3 and then in 1.0 cm^3 steps up to 35.0 cm^3. Measure the e.m.f. after each addition. Plot the e.m.f. against titre, determine the end point and so the molarity of the permanganate solution.

5.9 Gravimetric determination of lead

$$Pb^{2+} + CrO_4{}^{2-} \rightarrow PbCrO_4(s)$$

Dry a clean sintered glass crucible in an oven. Allow it to cool in a desiccator and weigh accurately.

Weigh out accurately about 0.3 g of lead(II) nitrate and dissolve the crystals in water (200 cm^3). Acidify the solution with 2 M ethanoic acid and then heat to boiling. Carefully add 4 per cent potassium chromate solution (a skin irritant) until the solution is yellow. Cover the beaker with a clock glass and boil gently for 15 minutes to complete the coagulation of the precipitate. If the solution loses its yellow colour, add a little more of the chromate.

Filter off the lead(II) chromate through the sintered glass crucible by gentle suction. Wash the residue with hot water. When the washings are colourless, dry the crucible in an oven at 120 °C, cool in a desiccator and reweigh. Repeat the heating and weighing until the masses are constant. Calculate the percentage of lead in the chromate (from the formula) and so the percentage in the nitrate.

5.10 Gravimetric determination of the sulphate

$$Ba^{2+} + SO_4{}^{2-} \rightarrow BaSO_4(s)$$

Dry a porcelain crucible and lid by heating in a flame; allow it to cool in a desiccator and weigh.

Weigh out accurately about 0.3 g of the sulphate (e.g. potassium sulphate), dissolve it in water (200 cm^3) and add concentrated hydrochloric acid (0.5 cm^3). Boil the solution and add barium chloride solution (12 cm^3 of 5 g dihydrate per 100 cm^3 water). Cover the beaker with a clock glass and heat on a steam bath for 1 hour. Test the supernatant liquid periodically with barium chloride (1–2 cm^3) to ensure complete precipitation. Do not add a large excess of the barium salt.

Filter the suspension through an ashless filter paper, wash with hot water and transfer the paper to the crucible. Heat gently to dry the paper, and then raise the temperature to ignite the paper (p. 139). Allow the crucible to cool in a desiccator and weigh it. Repeat the heating–weighing procedure until the crucible has attained a constant mass.

Determine the percentage of sulphate in $BaSO_4$, and so the percentage of sulphate in the original material.

5.11 Verification of the Beer–Lambert law

Prepare a 5.0 × 10^{-4} M solution of potassium permanganate in 2 M sulphuric acid.

Plot an absorbance curve for the region 400–720 nm, using a 10 mm cell and a 2 M sulphuric acid reference cell.

Prepare a series of permanganate solutions by dilution of the initial one, and record the spectra for each, on the same chart, without changing the instrumental settings. Choose the wavelength at the peak (about 506 nm) and determine the absorbance in each case. Plot the absorbance against the concentration. How does the graph relate to the Beer–Lambert law?

5.12 Calibration curve for quantitative infrared analysis

Prepare a series of solutions in the concentration range 1–10 per cent of naphthalene in cyclohexane.

Using a fixed path length cell, run a trace for pure cyclohexane in the range 650–850 cm^{-1}. Adjust the transmittance to obtain the maximum transmittance (minimum absorbance) in this range. Retaining these settings, run the spectra of the naphthalene solutions on one sheet of paper. Plot the absorbance of the naphthalene at the peak against the concentration. The graph can then be used to determine the concentration of another solution of naphthalene.

5.13 Determination of the composition of tap water by flame photometry

The flame photometer should be operated according to the manufacturer's instruction manual.

Prepare the following solutions as standards:

1.000 mg Na cm^{-3} : dissolve 0.635 5 g NaCl in 250.0 cm^3 solution;
1.000 mg K cm^{-3} : dissolve 0.477 3 g KCl in 250.0 cm^3 solution;
1.000 mg Ca cm^{-3} : dissolve 0.624 3 g CaCO$_3$ in a little dilute hydrochloric acid and make it up to 250.0 cm^3 solution with water.

From these solutions (containing 1 000 parts of the ion per million of solution), prepare standards containing 20, 10, 5, 2, 1 ppm. Run each of the Na solutions through the photometer to produce a calibration curve and then a sample of tap water. If the tap water is too concentrated in Na$^+$ ions to give a reading, dilute it by a factor of ten with deionised water.

Repeat the procedure for the other standards and so determine the composition of tap water in terms of Na, K and Ca ions.

References

An Introduction to Analytical Chemistry; G. F. Lewis (BDH, 1973).
A Textbook of Quantitative Inorganic Analysis; A. I. Vogel (Longman, 4th edn., 1978).
Modern Analytical Methods; D. Betteridge and W. E. Hallam (RIC monograph, 1972).
Principles of Titrimetric Analysis; E. E. Aynsley and A. B. Littlewood (RIC monograph, 1962).
Potentiometric Titrations (Pye Scientific Instruments, Application Sheet No. 8).
Electrometric Methods; D. R. Browning (McGraw-Hill, 1969).
Spectroscopy; D. R. Browning (McGraw-Hill, 1969).
A survey of molecular spectroscopy; J. H. J. Peet, *Sch. Sci. Rev.,* 1972, **54**, 281–98.

Spectroscopy in Chemistry; R. C. Whitfield (Longman, 1969).
Some applications of visible light absorption spectrophotometry; D. H.
Mansfield, *Sch. Sci. Rev.*, 1971, **53**, 350–3.

Films

Acid–base indicators (CHEM Study).
Titration of a weak acid (Encyclopaedia Britannica).
Oxidation–Reduction (McGraw-Hill).
Molecular spectroscopy (CHEM Study).
Infrared spectroscopy (Perkin–Elmer).

Questions

1. 25.00 cm^3 of a 0.100 M sodium carbonate solution required 24.00 cm^3
    hydrochloric acid on titration using phenolphthalein as indicator. 16.80 cm^3
    of the same acid were required for 25.00 cm^3 of sodium hydroxide solution.
    Calculate: (*a*) the molarity of the acid; (*b*) the molarity of the sodium
    hydroxide solution; (*c*) the concentration of alkali.

2. A specimen of iron wire (2.544 g) was dissolved in dilute sulphuric acid
    and made up to 500 cm^3 of solution. A 25.00 cm^3 portion of this solution
    required 22.60 cm^3 of a 0.020 M potassium permanganate solution. What
    is the percentage purity of the iron wire?

3. Magnesium chloride forms a crystalline solid of formula $MgCl_2 \cdot nH_2O$.
    From the following data determine the value of n. 2.54 g of the salt
    were dissolved in 250 cm^3 aqueous solution. A portion (25.00 cm^3) was
    titrated against silver nitrate solution (0.090 M) and required 27.80 cm^3.

4. The following results were obtained in the conductimetric titration of
    hydrochloric acid (0.100 M) against 25.00 cm^3 sodium hydroxide solution.

Volume of acid/cm^3	Conductance/ 10^{-2} ohms^{-1}	Volume of acid/cm^3	Conductance/ 10^{-2} ohms^{-1}
0.0	1.40	22.0	0.40
2.0	1.30	24.0	0.41
4.0	1.15	26.0	0.50
6.0	1.05	28.0	0.59
8.0	0.92	30.0	0.69
10.0	0.81	32.0	0.76
12.0	0.74	34.0	0.83
14.0	0.65	36.0	0.90
16.0	0.57	38.0	0.96
18.0	0.51	40.0	1.03
20.0	0.45		

Plot the results and calculate the molarity of the base.

5. A sample of a calcium salt (0.555 g) was dissolved in water and treated with ammonium ethan-1,2-dioate solution.

$$Ca^{2+} + (NH_4)_2 C_2 O_4 \rightarrow CaC_2 O_4 (s) + 2NH_4^+$$

The ethan-1,2-dioate was filtered off, dried and heated to a constant mass of calcium carbonate (0.500 g).

$$CaC_2 O_4 \rightarrow CaCO_3 + CO$$

Determine the percentage of calcium in the original salt.

6. 0.100 g steel was dissolved in acid to give 500 cm^3 of solution. After treatment with a suitable complexing agent to produce an intense colour, the solution was analysed colorimetrically for the presence of copper. An optical density of 0.41 was obtained. A series of standard solutions of copper gave the results indicated. What was the percentage of copper in the steel?

Concentration/ 10^{-3} g dm^{-3}	Optical density
0.20	0.17
0.40	0.34
0.60	0.51
0.80	0.68
1.00	0.85

Part 3

Inorganic chemistry

Chapter 6

Inorganic chemistry

Background reading

The background to this chapter is covered in *Fundamentals of Chemistry*, Chapters 8–11.

Periodicity

In his original design of the periodic table, Mendeleev arranged the elements so that the elements of similar chemical and physical properties were sited in the same vertical columns or groups. If they were also arranged in order of increasing atomic number, the rows or periods show a gradual transition in their properties; for example, ranging from reactive metals through less reactive metals and non-metals to the reactive halogens. Newlands had previously noted a similar characteristic. By arranging the elements in numerical order, he found that every eighth element was similar to the first (the noble gases were unknown to both of these workers). The phenomenon of gradual change and periodic repetition was described as **periodicity**. The modern periodic table has been redrawn to take into account the electronic structure of the elements and the discovery of the noble gases (Fig. 6.1), but the principle is basically unchanged.

The feature of periodicity is clearly demonstrated in a number of properties.

(*a*) Electronic configuration

The modern periodic table can be considered in four blocks (Fig. 6.2),

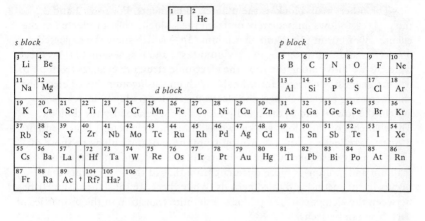

Fig. 6.1 Periodic table

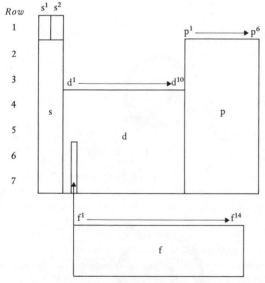

Fig. 6.2 Structure of the periodic table

related to the energy levels being filled with electrons. The elements of the s block are arranged in two columns, those with an s^1 configuration and those with an s^2 configuration. Hydrogen and helium nominally fall into this category, but, because of their size and the absence of any p orbitals, their chemistry is somewhat different and so they are usually dealt with separately.

The other 'main block' is the p block of elements. For rows 2 and 3, the p block follows immediately on from the s block. This is reflected in the chemistry of boron ($2p^1$) and aluminium ($3p^1$) which show the expected properties and transitions from beryllium ($2s^2$) and magnesium ($3s^2$) respectively. As we move along the row, the electronic structure changes from p^1 to p^6; the chemistry changes accordingly. With increasing numbers of electrons and increasing nuclear charge, the electronegativity gradually increases.

In the rows 4 and 5, the s and p blocks are separated by the d block. The elements of this block are known as 'transition elements', because their properties reflect a gradual transition from those of the s block to those of the p block. Consequently, the chemistry of the $4p^1$ and $5p^1$ elements (gallium and indium) is more different from that of the s^2 elements than occurs in the second and third periods. While there are substantial similarities between the elements of the p block, a definite transition in the properties of each row can be seen.

The electrons involved in the construction of a period (e.g. the 4s, 3d and 4p orbitals of the fourth period) are described as the valency electrons. These electrons are of importance not only in the bonding but also the stereochemistry of the compounds (p. 176).

The chemistry of the elements of each of these blocks is covered in the background reading material (see p. 154).

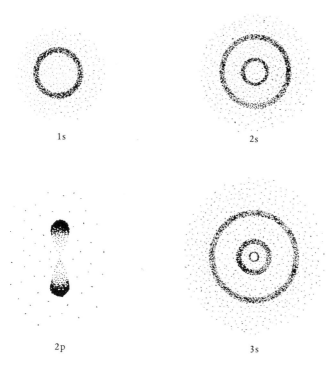

1s

2s

2p

3s

3p   3d

Fig. 6.3 Probability diagrams for hydrogen orbitals

(b) Radii

Sizes of atoms can be described on a number of bases and it is important to distinguish between these if unnecessary ambiguities are to be avoided.

The size of an unbonded atom is difficult to quantify, since the electronic structure (as interpreted currently) is rather diffuse in its radial nature (*Fundamentals of Chemistry,* pp. 55–6). The electron density of an atom varies with its atomic number, and each orbital has its own probability pattern (Fig. 6.3). Since these are described as a probability distribution, (*Fundamentals of Chemistry,* p. 19), there is no position beyond which the electron density is zero. It is, therefore, conventional to state the atomic size of an unbonded atom by the radius which encloses 90 per cent of the electron density. Figure 6.4 describes these radii for two atoms of differing complexity, hydrogen and rubidium.

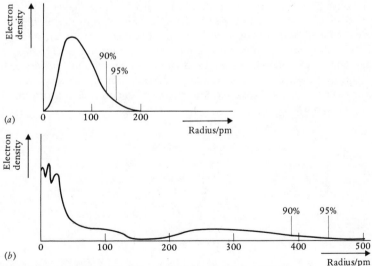

Fig. 6.4 Atomic radii of (*a*) hydrogen and (*b*) rubidium

An alternative definition of the atomic radius is that known as the **van der Waals' radius**. This is half the distance between the nuclei of two touching, but non-bonded, atoms of the element (Fig. 6.5). In simple non-bonded

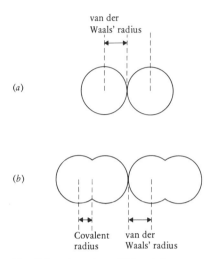

(a)

(b)

Fig. 6.5 Atomic radii for (a) monatomic and (b) diatomic molecules

atoms such as the noble gases, the van der Waals' radius is half the distance between the nuclei of any pair of touching atoms. If a diatomic molecule is formed, the distance under consideration is that between the nuclei of the touching, but non-bonded, atoms.

Covalent radii are defined as half the distance between singly bonded identical atoms. The ideal situation is that in a hydrogen molecule, H—H, or a halogen, Cl—Cl. Unfortunately, many diatomic molecules involve multiple bonding ($O{=}O$, $N{\equiv}N$). The degree of overlap of two atoms is dependent on the bond order (number of covalent bonds between two atoms). The bond length decreases as the bond order increases (see *Fundamentals of Chemistry*, p. 219). This is because the second, and subsequent, bonds require a contraction of the molecule to produce increased overlap of the orbitals constituting the π-bond (Fig. 6.6).

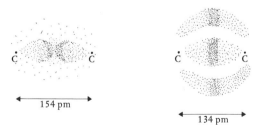

Fig. 6.6 Effects of multiple bonding on a bond length for carbon

The bond length is also affected by the electronic environment of the atoms, so, if the C—C bond length is required, a symmetrical molecule involving non-polar bonds (e.g. C—H) should be involved (e.g. an alkane). Therefore, the covalent radius must take into account the full chemical nature of a molecule.

For metals, the **metallic radius** can be defined as half the distance between the nuclei of the close-packed atoms (see p. 18).

Ionic radii are defined in a similar manner, being half the distance between two touching identical ions in a close-packed lattice (Fig. 6.7).

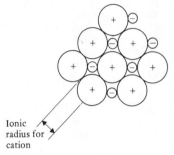

Ionic radius for cation

Fig. 6.7   Ionic radii

Unfortunately, this is also difficult to measure directly in many cases, since few ionic lattices are free of distortion. The values also vary according to the ionic charge.

Table 6.1 lists some of the typical values for these radii. From this table it will be appreciated that any consideration of the periodicity of size must involve a consistent set of figures (e.g. the covalent radii of singly bonded atoms). Many diagrams drawn in the past have used, for example, the covalent radii of most elements and the van der Waals' radii of the noble gases! Some erroneous conclusions have been drawn as a result. Figure 6.8 illustrates the variation in the covalent radii with atomic number. It will be observed that elements in similar relative positions in the figure (e.g. the maxima) are in the same groups of the periodic table (the alkali metals, s^1 group, in this case).

(c) Ionisation energies

The ionisation energy of an atom is determined by three main factors: (i) the distance of the electron from the nucleus and so the positive charge holding it to the atom; (ii) the magnitude of the nuclear charge; (iii) the shielding effect of the inner electrons.

Figure 6.8 shows that the size of an atom increases on descending a group of the periodic table. For example, consider the s^1 group (Li, Na, K, Rb, Cs and Fr) in the figure; they occupy peak positions in the curve, but the radius increases on passing from lithium to francium. It will also be seen that there is a gradual decrease in size on moving from the s to the p configurations. As a result, we find that the ionisation energy decreases on

Table 6.1 Some typical radii

Atom	Metallic radius/ pm	Covalent radius/ pm	Ionic radius (noble gas configuration)/ pm	van der Waals' radius/pm
Hydrogen		37	154	12
Helium				99
Lithium	152	123	68	
Beryllium	112	106	30	
Boron	79	88	16	
Carbon		77	16	
Nitrogen		70	171	150
Oxygen	60	66	146	140
Fluorine		64	133	135
Neon		71		160
Sodium	186	157	98	
Magnesium	160	140	65	
Aluminium	143	126	45	
Silicon	118	117	78	
Phosphorus		110	212	190
Sulphur	95	104	190	185
Chlorine		99	181	180
Argon		98		192
Potassium	227	203	133	
Calcium	197	174	94	
Scandium	161	144	81	
Titanium	145	13?	68	
Vanadium	132	1?2	59	
Chromium	137	117	52	
Manganese	137	117	46	
Iron	124	116		
Cobalt	125	116		
Nickel	125	115		
Copper	128	135	96	
Zinc	133	131	74	
Gallium	122	126	62	
Germanium		122	53	
Arsenic	125	118	222	200
Selenium		114	202	200
Bromine	115	111	196	195
Krypton		112		197

descending a column, because the attractive energy between the nucleus and the electron is diminished by the greater distance. As the nuclear charge increases, the distance decreases due to the greater force of attraction on the electron by the nucleus. Consequently, as we move across a row, we find that the size of the atom decreases, but the ionisation energy increases.

The effectiveness of the inner s, p and d orbitals in shielding the outer electrons from changes in the nuclear charge is different. The efficiency with which the different electronic orbitals shield the valence electrons decreases in the order s > p > d > f. If one considers the electronic arrangement of the elements, it is apparent that the influence of the nuclear charge (and so the

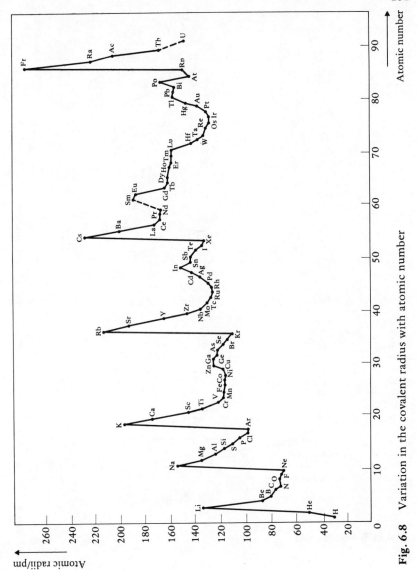

Fig. 6.8 Variation in the covalent radius with atomic number

ionisation energy) decreases in the order of valence electrons occupying the orbitals p (shielded by d), s (shielded by p) and d (shielded by s).

Figure 6.9 shows the variation in the ionisation energy with atomic number. Periodicity is displayed by the occurrence of similar elements at corresponding parts of the graph. For example, the group of noble gases all occur at maximum positions, having the highest ionisation energies in each period. Conversely, the alkali metals occupy the minimum positions.

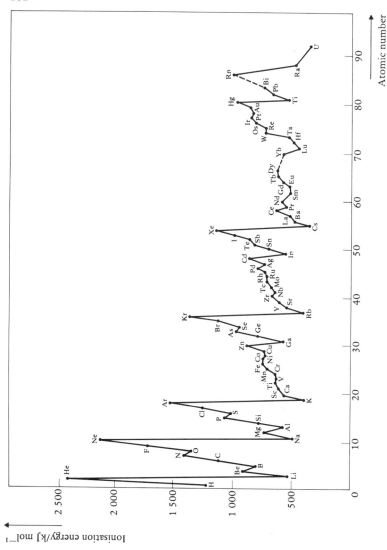

Fig. 6.9 Variation in the ionisation energy with atomic number

The factors outlined above can be illustrated from this figure too. Consider the row of elements potassium through to krypton. Potassium has a lower ionisation energy than sodium. This is due to an increase in the size of the potassium compared to sodium (point (i) above). On moving across the row from potassium, the second effect (ii) is observed, as the magnitude of the nuclear charge increases. (This effect is complemented by an accompanying slight decrease in size.) Shielding effects are also apparent. Since the d orbitals are protected by the s electrons (e.g. $4s^2$ for the 3d elements), there is only a minimal variation in their ionisation energies on passing along

163

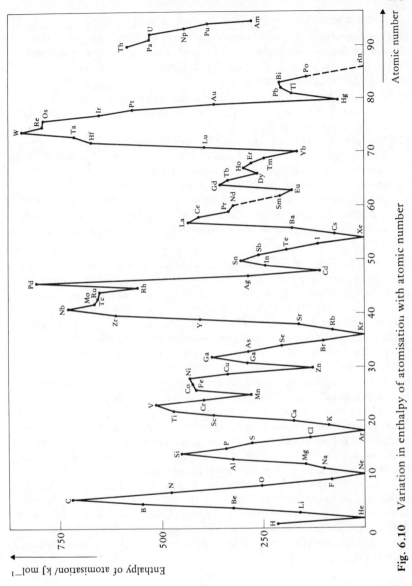

Fig. 6.10 Variation in enthalpy of atomisation with atomic number

the row. This is because the s electrons shield the d electrons from most of the nuclear charge increase. The 4p electrons in the elements Ga to Kr are less effectively protected (by the relatively poor shielding 3d electrons) and so there is a much larger effect by the nuclear charge. As a result, these elements show a large variation in their ionisation energies.

(d) Bond energies

There is a significant degree of periodicity too in the bond energies of the elements of the periodic table. This can be demonstrated in the enthalpies of atomisation of the elements (Fig. 6.10). There is a varying degree of bonding between the elements.

1. For some, there is little or no bonding, as, for example. in the noble gases in which there are only the weak interatomic forces (*Fundamentals of Chemistry*, pp. 50–2).
2. For other elements, the atoms are bonded in small units, such as in the diatomic gases O_2, H_2, Cl_2. In these, there is some variation in the bond energies, determined by factors such as the bond order (*Fundamentals of Chemistry*, p. 219) and atomic size.
3. Other elements involve three-dimensional networks in which each atom is affected by a number of others. Examples of these include the metals with their metallic bonds and non-metals such as diamond with strong covalent bonds.

There are also some intermediate structures involving units larger than those mentioned in (2), but not as extensive as those in (3). For example, black phosphorus (see Fig. 6.11) is two-dimensional in its structure; sulphur also has a layer structure. Figure 6.10 illustrates these differences and the associated trends. As we move across a row of the table, there is an increase in the number of valence electrons and an increase in electronegativity. Consider the row Na–Ar. On moving from sodium to silicon, there is an increasing number of valence electrons. If these were involved in metallic bonding, the bond energy would become more negative (i.e. more energy would be released) due to the increase in the number of electrons producing the metallic bond. But, there is another important factor to consider; there is an increase in electronegativity and so a decreasing tendency to lose electrons to form metallic bonds. So, in these four elements, the enthalpy of atomisation increases because of an increasing number of valence electrons, but the bond nature becomes increasingly covalent.

From silicon to argon, the enthalpy of atomisation decreases and this can be related to the fact that these covalent species form decreasing numbers of covalent bonds. Silicon forms four covalent bonds and so is able to form a three-dimensional structure; atomisation requires a breakdown of these four bonds. Phosphorus can only form three covalent bonds, having three unpaired electrons in its p orbitals:

As a result, phosphorus forms a volatile unit, P_4, in its white form, or the less volatile red and black forms. These involve larger molecular units (Fig. 6.11) but these are still based on three-bonded phosphorus atoms. Sulphur has two covalent bonds and most commonly occurs in S_8 units. Chlorine is diatomic

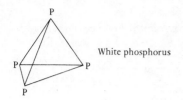

White phosphorus

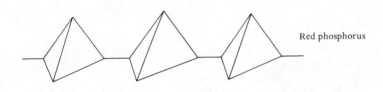

Red phosphorus

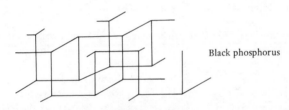

Black phosphorus

Fig. 6.11 Allotropes of phosphorus

and involves a single covalent bond. The noble gas, argon, has no unpaired electrons and shows no tendency to bond formation. It is, therefore, monatomic and requires no energy for atomisation.

Similar results are shown in compound bond energies. The term is not applicable to ionic compounds in the same way as to covalent compounds. For the latter, the bond energy is the energy change when two atoms are bonded together:

$$A^{\cdot} + {\cdot}B \rightarrow A{:}B$$

The equivalent term for ionic species is the energy change for the process

$$A^{+} + B^{-} \rightarrow A^{+}B^{-}$$

and this is known as the lattice energy. Complete sets of data for these changes are more difficult to obtain, but some values for the halides can be considered. Table 6.2 shows that, for a group of elements in the periodic

Table 6.2 Bond energies (in kJ mol^{-1}) for some halides

Element	Halide F	Cl	Br	I
Li	−1 029	−849	−804	−753
Na	−915	−781	−743	−699
K	−813	−710	−679	−643
Be	−3 456	−2 983	−2 895	−2 803
Mg	−2 883	−2 489	−2 414	−2 314
Ca	−2 582	−2 197	−2 125	−2 038
C (in CH_3X)	−452	−352	−293	−234

table, the bond energy of the halides increases (that is, it becomes less negative) as we descend the group. For example, the energy increases for each of the elements listed on moving from the fluorides to the iodides. The bond energy changes in the same manner for the metal halides, lithium to potassium and beryllium to calcium.

(e) Oxidation states

There is also a degree of periodicity in the range of oxidation states available. For example, in the hydrides of rows 2 and 3 (i.e. the lithium and sodium rows) the pattern of the oxidation states is shown in Table 6.3. Oxygen compounds increase the range of states available (Table 6.4).

Table 6.3 Oxidation states for hydrides of rows 2 and 3 of the periodic table

Configuration	s^1	s^2	p^1	p^2	p^3	p^4	p^5	p^6
Oxidation number	1	2	3	4	3	2	1	0
Examples	Li^+H^-	BeH_2	B_2H_6	CH_4	NH_3	H_2O	HF	Ne
	Na^+H^-	$Mg^{2+}H^-_2$	AlH_3	SiH_4	PH_3	H_2S	HCl	Ar

Table 6.4 Oxidation states for the oxides of row 2 of the periodic table

Configuration	s^1	s^2	p^1	p^2	p^3
Oxidation numbers and examples (max. and min.)	1, $Li^+_2O^{2-}$	2, $Be^{2+}O^{2-}$	3, B_2O_3	4, CO_2	5, N_2O_5
				2, CO	1, N_2O

Configuration	p^4	p^5	p^6
Oxidation numbers and examples (max. and min.)	6, SO_3	7, Cl_2O_7	8, XeO_4
	4, SO_2	1, Cl_2O	6, XeO_3

For these it will be observed that the maximum value increases across the row. The states listed are not necessarily available for each element in the group. This same feature may be observed in the hydroxy compounds, bases or acids, as in Table 6.5.

Table 6.5 Oxidation states for hydroxy compounds of the periodic table

Configuration	s^1	s^2	p^1	p^2	p^3	p^4	p^5	p^6
Oxidation state	1	2	3	4	5	6	7	8
Examples	Na^+OH^-	$Mg^{2+}(OH^-)_2$	H_3BO_3	H_2CO_3	HNO_3	H_2SO_4	$HClO_4$	H_4XeO_6
				2	3	4	5	6
				$Pb(OH)_2$	HNO_2	H_2SO_3	$HClO_3$	H_2XeO_4
							3	
							$HClO_2$	
							1	
							$HClO$	

Similar features occur in the longer rows of the table (e.g. row 4) which include the d-block elements. These elements include a more extensive range of oxidation states than are normally shown in the s and p blocks of the table.

There are, therefore, a number of features which vary in a regular manner related to the position of the element in the table. The variation occurs both across a row and down a group. These latter features are now examined in the specific cases of the groups p^2 and p^3.

The p² group

This group consists of the elements carbon, silicon, germanium, tin and lead. Table 6.6 lists some data for these elements. In accordance with the pattern seen earlier (p. 159), the covalent radii are observed to increase down

Table 6.6 Data on the p^2 elements

Element	Atomic number	Relative atomic mass	Covalent radius/pm	1st Ionisation energy/kJ mol^{-1}
Carbon	6	12.011	77	1 090
Silicon	14	28.086	117	790
Germanium	32	72.590	122	760
Tin	50	118.69	140	710
Lead	82	207.19	154	720

the group. As a consequence, the ionisation energies follow a decrease on moving from carbon to tin. Contrary to our predictions, lead has a larger value than expected. This can be related to the fact that it occurs in a row involving the 4f elements. Reference to page 160 shows that the f orbitals are relatively inefficient at shielding the electrons from the effects of the nuclear charge, and so the first valence electron of lead is less readily moved than that in tin despite the increased size of the atom.

Because of the changes in the ionisation energies, there is a corresponding change in the nature of the chemical bonding. Carbon exists in two non-metallic forms – graphite and diamond (Fig. 6.12). These two forms, both

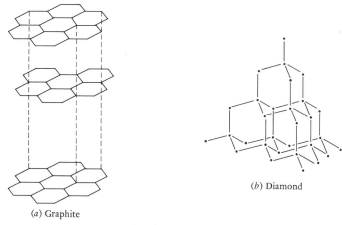

(a) Graphite

(b) Diamond

Fig. 6.12   Allotropes of carbon

pure carbon and in the same physical state, are called **allotropes** of carbon. The atoms are bonded together by covalent bonds. In diamond, the four bonds are orientated tetrahedrally around the central atom giving a three-dimensional structure.

Graphite, in contrast, is a layer structure. It has three covalent bonds on each carbon in one plane binding the atoms together. The fourth electron forms a π-bond with neighbouring carbon atoms in a manner similar to benzene (see *Fundamentals of Chemistry*, p. 218). This is illustrated in Fig. 6.13. The carbon atom 1 is bonded to atoms 2, 3 and 4 by σ-covalent bonds. The π-bonds of the rings A, B and C are interrelated, since the fourth electron of 1 can be shared with atoms 2 and 3 of ring A, with atoms 2 and 4 of ring B and with atoms 3 and 4 of ring C. These π-systems will be similarly linked with adjacent rings, spreading the π-electrons over the whole layer. The layers of atoms in graphite are only held together by the weak van der Waals' forces.

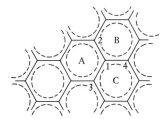

Fig. 6.13 π-bonding in graphite

As a consequence, diamond is hard and non-conducting, being made up of strong C—C bonds in all directions, and the bonding electrons are generally localised between the atoms preventing them from conducting electricity. On the other hand, the layers of graphite can move across each other with little hindrance (so it can act as a solid lubricant) and the molecule can conduct electricity along the plane of the layers (in which there is a mobility of the π-electrons), though not in a direction perpendicular to these layers.

Silicon and germanium have the diamond structure. They display an important property known as semiconduction. Semiconductors are important in the production of transistors. Germanium is a non-conductor at room temperature, but on heating the resistance falls. It is an **intrinsic semiconductor.** The orbitals involved in bonding in germanium are fully filled with electrons, but there are empty orbitals close to these. The increase in thermal energy causes electrons to pass from the filled orbitals to the empty ones. Because of the three-dimensional network of bonding in this diamond-type structure, the empty orbitals also interact. The excited electrons are able, therefore, to move across the lattice in a manner similar to that in metals (p. 14). An alternative form of semiconduction (**impurity semiconduction**) is displayed when a trace (1 part in 10^6) of a p^1 or a p^3 metal is introduced into the lattice. The resistance can be reduced by a factor of 50. Impurity semiconductors fall into two groups: N-type and P-type. The N-type semiconductors are produced by a p^3 element such as antimony. If antimony atoms replace some of the germanium atoms, the conductivity rises due to the extra electron ($s^2 p^3$ instead of $s^2 p^2$) having to occupy the formerly empty bands. The conductivity arises from the donation of negative charge. The introduction of aluminium (a p^1 element) causes a deficiency of electrons, corresponding to a net positive charge. These P-type conductors give rise to a flow of current through the 'holes' caused in the filled band due to this deficiency of electrons.

Tin exists as three allotropes; they are known as grey, white and brittle tin. Grey tin has a diamond-type structure, wherease white tin has a body-centred lattice. The brittle form assumes a close-packing arrangement. Lead has only the cubic close-packing arrangement. Both of these metals are electrical conductors with metallic bonding.

Chemical properties

It has been noted (p. 159) that the ionisation energies decrease on descending the group. As a result of the gradation in ionisation energies, there is an increase in the ionic character of the compounds of these elements. There is also a gradation in the chemical properties of the elements.

Carbon is relatively inert towards other chemicals. For example, graphite only reacts with oxygen above 960 K, giving carbon dioxide:

$$C(s) + O_2(g) \rightarrow CO_2(g), \Delta H = -395 \text{ kJ mol}^{-1}$$

Diamond requires higher temperatures before significant reaction occurs. Graphite reacts with concentrated oxidising acids to form a yellow-brown substance of unknown composition, known as graphite oxide.

Silicon is significantly more reactive. It burns in oxygen at a lower temperature (670 K) than graphite:

$$Si(s) + O_2(g) \rightarrow SiO_2(s), \Delta H = -702 \text{ kJ mol}^{-1}$$

It reacts with halogens, e.g. chlorine at 500 K, with sulphur vapour, and with nitrogen (at 1600 K). It is unattacked by acids, but reacts with concentrated alkali to form a silicate. Steam attacks silicon at red heat.

Germanium has a similar range of reactions. It reacts with oxygen, halogens, sulphur and alkali. Though it is unattacked by non-oxidising acids, concentrated nitric acid precipitates germanium(IV) oxide. Each of these three elements gives products in which the element is in the +4 oxidation state.

Tin forms tin(IV) compounds with oxygen, halogens, sulphur and concentrated nitric acid. The latter reactant produces tin(IV) oxide. Hot concentrated hydrochloric acid gives tin(II) chloride and hydrogen:

$$Sn + 2HCl \rightarrow SnCl_2 + H_2$$

Tin(II) sulphate and sulphur dioxide are formed in the reaction between tin and concentrated sulphuric acid:

$$Sn + 2H_2SO_4 \rightarrow SnSO_4 + SO_2 + 2H_2O$$

In addition to the increase in the reactivity of tin, compared to the earlier elements of this group, it will be noted that tin is only oxidised to the +2 state with the latter two acids.

Lead displays a continuing increase in reactivity together with a decreased tendency to the +4 state. Oxygen produces lead(II) oxide, PbO, and lead(II) lead(IV) oxide ('red lead'), Pb_3O_4; sulphur and the halogens also react with lead, giving the lead(II) species. Concentrated nitric acid generates lead(II) nitrate (contrast SnO_2 for tin). Though the cold hydrochloric and sulphuric acids show little effect on lead, the hot acids react as for tin giving the lead(II) chloride and sulphate respectively. The inactivity in the cold may well be due to the insolubility of the products in cold aqueous media. The only stable lead(IV) compound is the oxide PbO_2, which is prepared by hypochlorite oxidation of aqueous solutions of lead(II) salts:

$$Pb^{2+}(aq) + 2OCl^-(aq) \rightarrow PbO_2(s) + 2Cl^-(aq)$$

Oxidation–reduction properties

The possible interconversion of the oxidation states of an element gives a source of oxidants and reductants. It has been shown that in their normal chemistry carbon, silicon and germanium exist in the +4 state, tin exists in +2 and +4 forms, and lead is usually only prepared in the +2 state. The result of this is that lead(IV) oxide is an oxidising agent, reacting, for example, with hydrochloric acid to give lead(II) chloride and chlorine:

$$PbO_2 + 4HCl \rightarrow PbCl_2 + Cl_2 + 2H_2O$$

Lead(IV) is such a strong oxidising agent that it reacts with bromide and iodide ions, preventing the preparation of the lead(IV) halides. Tin produces a range of compounds in both states and the tin(II) compounds are reducing agents. For example, tin(II) chloride reduces iron(III) chloride to iron(II) chloride:

$$SnCl_2\,(aq) + 2FeCl_3\,(aq) \rightarrow SnCl_4\,(aq) + 2FeCl_2\,(aq)$$

Carbon forms an oxide in the lower state, carbon monoxide (CO). This can be produced in a number of ways: combustion of carbon in a limited amount of oxygen,

$$C(s) + \frac{1}{2}O_2\,(g) \rightarrow CO(g)$$

reduction of carbon dioxide with excess carbon,

$$CO_2\,(g) + C(s) \rightarrow 2CO(g)$$

and dehydration of methanoic acid by hot concentrated sulphuric acid,

$$HCOOH(l) \rightarrow CO(g) + H_2O(g)$$

This oxide is a good reducing agent and reduces many metal oxides to the metal. It is used commercially for the reduction of, for example, haematite (Fe_2O_3)

$$Fe_2O_3\,(s) + 3CO(g) \rightarrow 2Fe(s) + 3CO_2\,(g)$$

and zinc oxide

$$ZnO(s) + CO(g) \rightarrow Zn(s) + CO_2\,(g)$$

Hydrolysis of the halides

The chlorides of this group are covalent. A characteristic of covalent chlorides is their ready hydrolysis (see *Fundamentals of Chemistry*, p. 119). The process of hydrolysis may be illustrated with the case of silicon tetrachloride, $SiCl_4$. A polar water molecule attacks the silicon atom (which is $\delta+$ due to the higher electronegativity of the chlorine), using the available 3d orbitals as a means of attack:

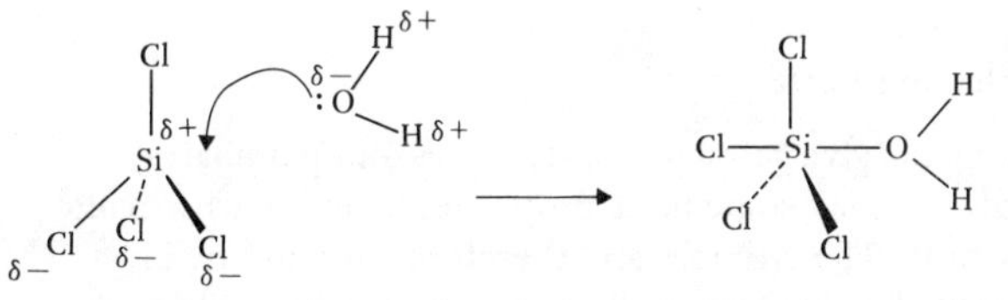

An internal reaction occurs to eliminate HCl:

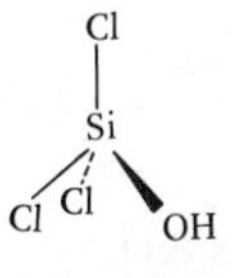

Attack by other water molecules causes replacement of each of the chlorine atoms to give '$Si(OH)_4$'. Analysis of the product indicates that it is best considered as hydrated silicon(IV) oxide, $SiO_2 \cdot 2H_2O$.

Germanium, tin and lead tetrahalides react similarly:

$$GeCl_4 + 2H_2O \rightarrow GeO_2 + 4HCl$$
$$SnCl_4 + 2H_2O \rightarrow SnO_2 + 4HCl$$
$$PbCl_4 + 2H_2O \rightarrow PbO_2 + 4HCl$$

The dihalides of tin and lead do not undergo such extensive hydrolysis and may be isolated as crystalline hydrates from aqueous solution, e.g. $SnCl_2 \cdot 2H_2O$.

The carbon halides do not undergo hydrolysis. This is probably due to a number of factors: the lack of available orbitals (there are no 2d orbitals), the strength of the carbon–halogen bonds and steric hindrance. The halogen atoms are so large that they prevent water molecules reaching the carbon atom (Fig. 6.14).

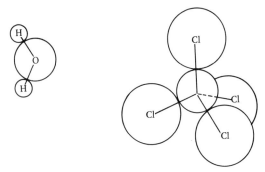

Fig. 6.14   Relative sizes of water and tetrachloromethane (carbon tetra-chloride)

The acid–base character of the oxides

All the dioxides display acidic character and react with bases to form salts:

$$CO_2(g) + 2NaOH(aq) \rightarrow Na_2CO_3(aq) + H_2O(l)$$
$$\text{sodium carbonate}$$
$$SiO_2(s) + CaO(s) \rightarrow CaSiO_3(s)$$
$$\text{calcium silicate}$$

Similarly, the other elements give germanates, stannates and plumbates. These three elements also display some basic properties, reacting with acidic substances to form the salts. These oxides are, therefore, amphoteric in nature. This decrease in acidic character is illustrated by the pK_a values of the early members:

$$CO_2 + H_2O \rightleftharpoons H^+(aq) + HCO_3^-(aq) \quad pK_a = 6.4$$
$$H_2SiO_3 \rightleftharpoons H^+(aq) + HSiO_3^-(aq) \quad pK_a = 9.9$$

The silicates occur extensively in rocks and clays, the basic units SiO_3^{2-} or SiO_4^{4-} forming more complex units (Fig. 6.15). This property is not displayed by the other anions to any significant extent.

Silicon does not form a monoxide. Carbon monoxide dissolves slightly in water without reaction (2.14×10^{-4} mol dm^{-3}) and so is classified as a neutral oxide. However, it does react with sodium hydroxide at high temperature and pressure to form the sodium salt of methanoic acid:

$$CO + NaOH \rightarrow Na^+(HCOO^-)$$

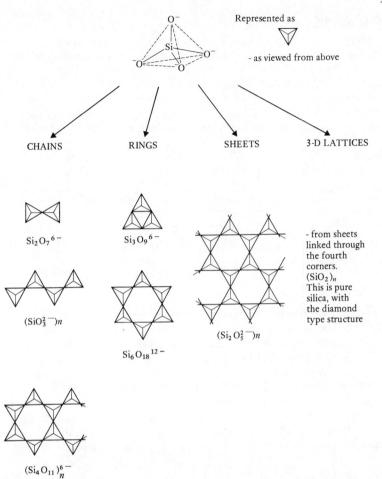

Fig. 6.15 Classes of silicate

Germanium(II) oxide, one of the rare compounds of the germanium(II) state, has a similar solubility to carbon monoxide (2×10^{-4} mol dm^{-3}). It is amphoteric forming the GeCl$_3$$^-$ ion with concentrated hydrochloric acid and GeO$_2$$^{2-}$ (germanates(II)) with sodium hydroxide solution.

Tin forms an amphoteric oxide of substantially lower solubility (5×10^{-6} mol dm^{-3}). With hydrochloric acid, the tin(II) chloride is formed; sodium hydroxide solution gives a stannate(II) in solution (Na$_2$SnO$_2$).

Lead monoxide, PbO, is predominantly basic, but it does display some amphoteric character. It has a low solubility, 1.08×10^{-4} mol dm^{-3}. The chloride and sulphate of lead are insoluble in cold water, though very soluble in hot water, and so the oxide only shows significant reaction with hot hydrochloric and sulphuric acids. The oxide will also dissolve in very concentrated sodium hydroxide to give a plumbate(II), PbO$_2$$^{2-}$.

The p³ group

This group (Table 6.7) is composed of essentially non-metallic elements. The lower members show partial metallic character and are often called

Table 6.7   Data for the p³ elements

Element	Atomic number	Relative atomic mass	Covalent radius/pm	1st Ionisation energy/kJ mol⁻¹
Nitrogen	7	14.006 7	74	$-1\,400$
Phosphorus	15	30.973 8	110	$-1\,010$
Arsenic	33	74.921 6	121	-950
Antimony	51	121.755 0	141	-830
Bismuth	83	208.980 0	152	-700

metalloids. Nitrogen is a diatomic gas in which the nitrogen atoms are bound by a σ- and two π-bonds:

$$N \equiv N$$

Phosphorus exists in a number of allotropic forms (see p. 164). 'White phosphorus' consists of tetrahedral P_4 units. The other common form of phosphorus is known as 'red phosphorus' and appears to be a polymerised form of the P_4 unit in which one bond of the tetrahedron is broken to link with other units (Fig. 6.11).

Red phosphorus is much less reactive than the white variety. For example, the white allotrope smoulders in air at ordinary laboratory temperatures (18–20 °C), though it ignites on warming, to give the oxide, P_4O_{10}; the red phosphorus only reacts after ignition in oxygen at higher temperatures. Another allotrope, black phosphorus, is formed at higher temperatures and pressures and is chemically inert. In this allotrope the other bridging group is broken to form a layer structure (Fig. 6.11).

Arsenic and antimony have yellow non-metallic forms with a tetrahedral structure (As_4 and Sb_4) analogous to white phosphorus. They also form the less reactive layer structures. Bismuth has only a metallic, layer form.

Chemical properties

Nitrogen is unreactive towards most common reagents. This can be related to the large bond dissociation energy:

$$N_2(g) \rightarrow 2N(g) \quad \Delta H = +945 \text{ kJ mol}^{-1}$$

Any reaction involving nitrogen must, therefore, compensate for this high energy requirement. Only lithium reacts at moderate temperatures, forming Li_3N. It does react with burning magnesium to form the nitride, Mg_3N_2. A few other elements react at red heat, for example, calcium, boron and aluminium. Reactions with hydrogen and with oxygen are used for the synthesis of ammonia

$$\tfrac{1}{2}N_2\,(g) + 1\tfrac{1}{2}H_2 \rightleftharpoons NH_3 \quad \Delta H_{298\,K} = -46 \text{ kJ mol}^{-1}$$

and nitrogen(II) oxide

$$\tfrac{1}{2}N_2 + \tfrac{1}{2}O_2 \rightleftharpoons NO \quad \Delta H_{298K} = +90 \text{ kJ mol}^{-1}$$

The reaction with hydrogen requires a low temperature to produce reasonable yields of product, but it is then too slow (see p. 56). Nitrogen(II) oxide is prepared by this route at high temperatures only; a more practical route involves the catalytic oxidation of ammonia:

$$4NH_3 + 5O_2 \underset{1100-1200 \text{ K}}{\overset{Pt - Rh}{\rightleftharpoons}} 4NO + 6H_2O$$

The oxide combines with oxygen and water to produce nitric acid.

The single bond in phosphorus is more easily broken than the triple bond of nitrogen:

$$\tfrac{1}{2}N_2 \rightarrow N \quad \Delta H = +473 \text{ kJ mol}^{-1}$$

$$\tfrac{1}{4}P_4 \rightarrow P \quad \Delta H = +316 \text{ kJ mol}^{-1}$$

The less reactive red allotrope has a value of 334 kJ mol^{-1}. This difference in the bond energies of the two elements is reflected in the increased reactivity of phosphorus. It reacts with oxygen at moderate temperatures to give the oxides P_4O_6 and P_4O_{10}, with very electropositive metals to give phosphides, with the halogens to give the tri- or pentahalides, and with sulphur to form a sulphide, P_2S_3. It is oxidised by concentrated nitric acid to orthophosphoric acid, H_3PO_4. Alkali attacks white phosphorus to give phosphine, PH_3. Because of its ready reaction with oxygen, *white* phosphorus is stored under water.

The lower elements of this group display similar chemical properties to those of phosphorus. They burn in oxygen and react readily with halogens and sulphur. They also react with metals. With concentrated nitric acid, there is a distinction in that arsenic gives the acid, H_3AsO_4, antimony gives the oxide, Sb_4O_6, and bismuth gives the nitrate, $Bi(NO_3)_3$. In this reaction we observe the trend to the lower oxidation states in antimony and bismuth (+3, in contrast to +5 in phosphorus and arsenic); also bismuth shows the increase in metallic character in the formation of a salt. Arsenic, like phosphorus, reacts with concentrated alkali, though in this case with the evolution of hydrogen.

The chlorides

A range of chlorides are known (Table 6.8). The non-existence of the +5

Table 6.8   The chlorides of the p^3 elements

NCl_3	PCl_3	$AsCl_3$	$SbCl_3$	$BiCl_3$
	PCl_5		$SbCl_5$	

state for nitrogen can be explained by the fact that there are no d orbitals available for use in bonding. For example, phosphorus has the structure $s^2 p^3$:

Clearly it can form three covalent bonds using the three unpaired electrons, hence PCl_3. The two s electrons may be separated and one raised to a higher energy level if sufficient energy is supplied (the 3d orbital is normally postulated here):

These five unpaired electrons can now bond with the halogens (so PCl_5), the geometry of the molecule being determined by hybridisation theory (see *Fundamentals of Chemistry,* p. 25). (There is some valid objection to the use of hybridisation theory, but it is a convenient model to use at this level.) Nitrogen does not form this structure because of the absence of d orbitals. Antimony follows phosphorus in the use of this +5 state, but, surprisingly, arsenic only displays it in its fluoride and oxide. In keeping with the trend noticed in the p^2 group, the lowest element, bismuth, displays only the lowest oxidation state.

All the chlorides undergo hydrolysis. There are three types of compound formed: (*a*) nitrogen trichloride gives the base, ammonia; (*b*) phosphorus and arsenic chlorides give the corresponding acids; (*c*) antimony and bismuth give the basic chlorides.

The hydrolysis to the acids involves the same procedure as that outlined above for silicon (p. 171):

$$XCl_3 + 3H_2O \rightarrow H_3XO_3 + 3HCl \; (X = P, As)$$

The pentahalides give the acids involving the +5 oxidation state:

$$PCl_5 + 4H_2O \rightarrow H_3PO_4 + 5HCl$$

This group of halides is hydrolysed through the initial coordination of water molecules to the central element; for example, in the case of phosphorus trichloride:

$$H_2O + PCl_3 \;\rightarrow\; H_2O \rightarrow PCl_3 \;\rightarrow\; (HO)PCl_2 + HCl$$

$$\downarrow 2H_2O$$

$$(HO)_3P + 2HCl$$

Antimony and bismuth are less electronegative and so do not undergo such complete reaction. For example, bismuth trichloride is reversibly hydrolysed to bismuth oxochloride:

$$BiCl_3 + H_2O \rightleftharpoons BiOCl + 2HCl$$

Nitrogen trichloride cannot be hydrolysed by the same mechanism because of the unavailability of suitable orbitals. The water molecules form a hydrogen-bonded species with the halide:

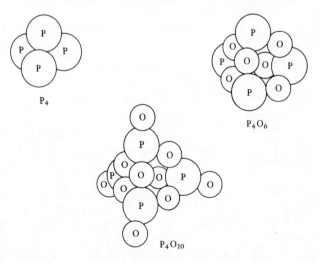

The oxides

Table 6.9 lists the main oxides of these elements. The common oxidation

Table 6.9 The oxides of the p^3 elements

Oxidation state				
+1	N_2O			
+2	NO			
+3	N_2O_3	P_4O_6	As_4O_6	Sb_4O_6	Bi_2O_3
+4	NO_2/N_2O_4			
+5	N_2O_5	P_4O_{10}	As_2O_5	Sb_2O_5	?

states are +3 and +5. The oxides of +3 phosphorus, arsenic and antimony have structures related to the elemental structures (Fig. 6.16). The nitrogen

P_4

P_4O_6

P_4O_{10}

Fig. 6.16 Structures of phosphorus and its oxides

(III) oxide has a distinct structure:

$$O = N - O - N = O$$

and is unstable both as a gas and in aqueous solution:

$$N_2O_3\,(g) \rightarrow NO(g) + NO_2\,(g)$$

$$N_2O_3 + H_2O \rightarrow 2HNO_2\,(aq) \rightarrow H_2O + NO + NO_2$$

Bismuth(III) oxide has a more complex structure but is predominantly ionic. While the oxides of nitrogen, phosphorus and arsenic are distinctly acidic, the antimony(III) oxide is amphoteric. For example, phosphorus(III) oxide forms phosphorous acid with water

$$P_4O_6 + 6H_2O \rightarrow 4H_3PO_3$$

arsenic(III) oxide reacts with concentrated base to give arsenates(III), AsO_3^{3-}; antimony(III) oxide reacts with hydrochloric acid to give the trichloride

$$Sb_2O_3 + 6HCl \rightarrow 2SbCl_3 + 3H_2O$$

and with bases to give antimonates(III):

$$Sb_2O_3 + 6OH^- \rightarrow 2SbO_3^{3-} + 3H_2O$$

Bismuth(III) oxide is entirely basic in its character, being insoluble in bases.

The oxides of the elements in the +5 state follow similar trends. The oxides decrease in stability from phosphorus to bismuth (compare the similar trend for the higher oxides of the p^2 elements). While the phosphorus(V) oxide is a stable, non-oxidising compound, the corresponding oxides of arsenic and antimony are oxidising agents (so being reduced to the +3 state), and the bismuth(V) oxide has not been isolated; its existence is implied by the formation of a bismuthate(V) on fusion of the bismuth(III) oxide and sodium hydroxide:

$$Bi_2O_3 + 2NaOH + \frac{1}{2}O_2 \rightarrow 2NaBiO_3$$

The bismuthate(V) is also an oxidising agent, oxidising, for example, manganese(II) ions to permanganate:

$$5BiO_3^-(s) + 2Mn^{2+}(aq) + 14H^+ \rightarrow 5Bi^{3+}(aq) + 2MnO_4^-(aq) + 7H_2O$$

The trend in acid–base character is also apparent:

P_4O_{10} – acidic
As_2O_5 – amphoteric
Sb_2O_5 – amphoteric

The arsenic(V) oxide dissolves in alkali to give the tetrahedral ion AsO_4^{3-}; antimony(V) oxide forms the octahedral ion $[Sb(OH)_6]^-$. The greater number of coordinated oxygens in the antimony ion reflects the increased size of this atom over arsenic. The bismuth oxide is unknown, but its oxo-salts are known (see above). Nitrogen also forms an oxide in the +5 state, N_2O_5. This is formed by the dehydration of nitric acid:

$$2HNO_3 \xrightarrow{P_4O_{10}} N_2O_5 + H_2O$$

It is a strong oxidising agent, indicating the instability of the higher state for this element. The oxide is formed by coordination of nitrogen to oxygen:

The phosphorus(V) oxide is of a similar structure to its lower oxide, with extra coordinated oxygens (Fig. 6.16), but the structures of the oxides of arsenic(V) and antimony(V) are unknown.

Nitrogen forms a number of other oxides too. Nitrogen(I) oxide can be obtained by the thermal decomposition of ammonium nitrate:

$$NH_4NO_3(s) \rightarrow N_2O(g) + 2H_2O(g)$$

It is a colourless, neutral oxide. It decomposes at high temperatures:

$$2N_2O \rightarrow 2N_2 + O_2$$

The action of copper on dilute nitric acid generates nitrogen(II) oxide.

$$3Cu(s) + 2NO_3^-(aq) + 8H^+(aq) \rightarrow 3Cu^{2+}(aq) + 2NO(g) + 4H_2O(l)$$

This is also a colourless, neutral oxide. It decomposes at high temperatures:

$$2NO \rightarrow N_2 + O_2$$

and combines with oxygen at ordinary temperatures:

$$2NO + O_2 \rightarrow 2NO_2$$

The product of this reaction, nitrogen(IV) oxide is also obtained by the thermal decomposition of nitrates (except those of the s^1 metals)

$$4NO_3^-(s) \rightarrow 2O^{2-}(s) + 4NO_2(g) + O_2(g)$$

and by the action of copper on concentrated nitric acid:

$$Cu(s) + 2NO_3^-(aq) + 4H^+(aq) \rightarrow Cu^{2+}(aq) + 2NO_2(g) + 2H_2O(l)$$

This brown gas dissolves in alkalis to form salts involving both the +3 (nitrates(III) or nitrites) and +5 (nitrates) oxidation states.

$$2NO_2 + 2OH^- \rightarrow NO_2^- + NO_3^- + H_2O$$

It is also an oxidising agent.

In summary it can be noted that for the p-block elements, on descending a group:

(*a*) metallic characteristics, such as ionic bonding, increase;
(*b*) the higher oxidation states decrease in stability;
(*c*) the lower oxidation states increase in stability;
(*d*) the oxides become less acidic and more basic;
(*e*) the chlorides show a decreasing tendency to hydrolysis.

180

Summary

At the conclusion of this chapter, you should be able to:

1. describe the periodicity of the elements displayed by their electronic structures, radii, ionisation energies, bond energies and oxidation states;
2. distinguish between the atomic, van der Waals', covalent, ionic and metallic radii;
3. describe the elemental forms of the p^2 elements and relate these to their physical and chemical properties;
4. describe the reactions of the p^2 elements with oxygen, halogens and acids;
5. describe the redox properties of carbon monoxide, lead(IV) oxide and of the tin(II) and (IV) states;
6. describe the hydrolysis reactions of the chlorides of the p^2 elements;
7. describe the acid–base character of the oxides of the p^2 elements;
8. describe the elemental forms of the p^3 elements;
9. describe the chemical reactions of the p^3 elements with hydrogen, oxygen, halides, acids and bases;
10. account for the formation of the +5 state in the phosphorus halide chemistry;
11. describe and account for the hydrolytic reactions of the p^3 chlorides;
12. describe the acid–base character of the oxides of the p^3 elements;
13. describe the preparation of the simple binary compounds and salts of the p^2 and p^3 elements;
14. summarise and illustrate the trends observed in the groups of the p block.

Experiments

6.1 Properties of the compounds of the p^2 elements

(a) Chlorides

1. Use the +4 state chlorides of these elements (C to Pb). Many of these are very reactive – they should he handled with DRY apparatus in a fume cupboard. Safety spectacles must be worn. To a few drops of each chloride in turn, add a little water dropwise. Note any changes observed. Test the product with universal indicator. Comment on the results.
2. Using such +2 state chlorides as are available, repeat the hydrolysis test. Account for the absence of certain chlorides and compare the behaviour of the two series of chlorides.

(b) Oxides

1. Determine the relative solubility in water (and pH of the solutions) of the dioxides of this group. For CO_2, bubble the gas through water.
2. Test the reactivity of the tin(IV) and lead(IV) oxides with hydrochloric acid and sodium hydroxide.
3. Repeat the previous tests using the monoxides.

(c) Oxoanions
1. Investigate the action of heat on a range of the oxosalts (carbonates, silicates, etc.) using a variety of cations (e.g. Ca^{2+}, Na^+, Zn^{2+}, Cu^{2+}).
2. Test the oxosalts with dilute hydrochloric acid.

(d) Cations
1. Compare the behaviour of the +2 and +4 ions of tin and lead (solutions of the chlorides in hydrochloric acid are suitable) with hydrogen sulphide. Comment on the results.
2. Add a little iron(III) chloride (2 drops) to aqueous solution (1 cm^3) of Sn^{2+} and Pb^{2+}. Add potassium hexacyanoferrate(III) (1 drop) to the product. This latter reagent gives a blue solution or precipitate with Fe^{2+} ions. Comment on the results.

6.2 Preparation of some +4 compounds of tin and lead

(a) Tin(IV) oxide

Place some granulated tin (1 g) and concentrated nitric acid in a beaker in a fume cupboard. If no reaction occurs, add 1 drop of water to initiate it. When the reaction has subsided and no tin remains, dilute the mixture with water and boil for 5 minutes. Filter off the hydrated oxide and heat it to dryness in a crucible.

(b) Tin(IV) chloride

Set up a standard distillation apparatus (Fig. 4.19) in a fume cupboard, and fit a dropping funnel to the flask and a calcium chloride tube (to protect the product from moisture) to the outlet tube. The apparatus must be dry (why?).

Place tin (1.2 g) in the flask and chlorosulphonic acid (3 cm^3) in the funnel. CARE must be exercised in handling this acid; it is very corrosive. Allow the acid to drip on to the metal very slowly. The initial reaction is very vigorous, but the reaction moderates as the tin is consumed. The product is purified by distillation (b.p. 113–115 $^\circ$C).

$$Sn + 4Cl \cdot SO_3H \rightarrow SnCl_4 + 2H_2SO_4 + 2SO_2$$

(c) Lead(IV) chloride

Cool a conical flask in ice and introduce lead(IV) oxide (1 g). Slowly add concentrated hydrochloric acid in sufficient quantity to just dissolve the oxide. Add cold ammonium chloride solution and filter off the ammonium hexachloroplumbate(IV). To this residue, add *carefully*, concentrated sulphuric acid. The lead(IV) chloride is produced as an oil.

6.3 Reactions of the compounds of the p^3 elements

(a) Reactions of the oxoanions(V)

Use the sodium salts of the nitrate, phosphate, arsenate, antimonate and bismuthate. Take great care with the arsenate and antimonate which are very poisonous.
1. Examine the effect of heat on small samples of these salts. Test for oxygen.

2. Warm solutions of the salts with ammonium molybdate solution.
3. Test solutions of the salts with silver nitrate solution; then acidify with dilute nitric acid.
4. Add small quantities of the solids to manganese(II) ions in solution. Warm.

(b) Reactions of the oxoanions(III)

Use the sodium salts of such anions as are available.
1. Test solutions of the salts with a little acidified permanganate solution.
2. Examine the effect of dilute hydrochloric acid on the salts.
3. Examine the effect of the solutions on copper(II) sulphate solution.

(c) Reactions of the chlorides

Examine the behaviour of the chlorides with water. Why is no NCl_3 or NCl_5 supplied?

References

The Periodic Table; D. G. Cooper (Butterworths, 4th edn., 1968).

Films

Carbon (Morgan Crucible Co. Ltd.).
Ammonia (ICI Ltd.).
Nitric acid (CHEM Study).
Chemistry and a changing world (Encyclopaedia Britannica) – phosphorus.
Lead – the enduring metal (Lead Development Association).

Questions

1. Using the information given on pages 154–167 concerning periodicity, predict as much as you can concerning the chemistry of gallium (Ga).
2. Write an account of the chemistry of silicon, with particular reference to its elemental properties, its oxide and chloride.
3. Describe the electrical properties of carbon, germanium and lead.
4. What are the products of the following reactions; give equations in each case?
   (a) tin + oxygen;
   (b) tin + concentrated sulphuric acid;
   (c) tin(II) sulphate and iron(III) chloride solutions;
   (d) tin(IV) chloride + water;
   (e) tin(II) oxide + sodium hydroxide solution.
5. Describe the industrial preparation of nitric acid, relating the conditions used to those predicted from the equilibrium:

$$4NH_3 + 5O_2 \rightleftharpoons 4NO + 6H_2O \qquad \Delta H = -907 \text{ kJ mol}^{-1}$$

Part 4

Organic chemistry

Chapter 7

Organic chemistry – a survey by functional groups

Background reading

The background material to this chapter is covered in *Fundamentals of Chemistry*, Chapters 15 and 16.

Reaction types

The important types of reaction of organic compounds, with typical examples, are summarised as follows.

(a) Addition reactions

Hydrogen bromide reacts with ethene to give bromoethane:

$$CH_2 = CH_2 + HBr \rightarrow CH_3 CH_2 Br$$

The addition of hydrogen cyanide to aldehydes is a useful means of extending the carbon chain. For example, with ethanal

$$CH_3 - \overset{\overset{\displaystyle O}{\|}}{C}\diagdown_H \ + \ HCN \rightarrow CH_3 - \overset{\overset{\displaystyle OH}{|}}{\underset{\underset{\displaystyle H}{|}}{C}} - CN$$

2-hydroxypropanenitrile

The reaction of an aldehyde or ketone with sodium hydrogen sulphite gives an ionic addition compound which is useful for the separation of aldehydes from impurities:

$$CH_3-\overset{\overset{\displaystyle O}{\|}}{C}-CH_3 + Na^+HSO_3^- \rightarrow CH_3-\underset{\underset{\displaystyle SO_3^-Na^+}{|}}{\overset{\overset{\displaystyle OH}{|}}{C}}-CH_3$$

(*b*) Substitution reactions

One atom or group of atoms can replace another to form an alternative compound. Halogenoalkanes can be prepared by the reaction of covalent chlorides and alcohols:

$$CH_3CH_2OH + HCl \xrightarrow{ZnCl_2} CH_3CH_2Cl + H_2O$$

Similarly, the displacement of halogens by cyanide is a useful route to the preparation of acids:

$$CH_3CH_2Cl + K^+CN^-(aq) \rightarrow CH_3CH_2CN + K^+Cl^-(aq)$$

$$CH_3CH_2CN + 2H_2O + HCl(aq) \rightarrow CH_3CH_2COOH + NH_4^+Cl^-(aq)$$

The amino group can be displaced by a hydroxy group using nitrous acid:

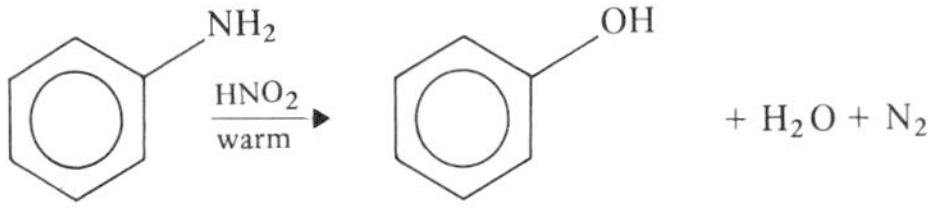

(*c*) Condensation reactions

Carbon–nitrogen double bonds can be introduced by means of the condensation reactions of carbonyl compounds:

$$(CH_3)_2C{=}O + H_2N\cdot OH \rightarrow (CH_3)_2C{=}N{-}OH + H_2O$$

Hydroxyammonia — Oximes

$$(CH_3)_2C{=}O + C_6H_5NH_2 \rightarrow (CH_3)_2C{=}N{-}C_6H_5 + H_2O$$

Anils

(*d*) Oxidation reactions

Aldehydes, ketones and carboxylic acids can all be obtained by the oxidation of alcohols:

$$CH_3OH \rightarrow HCHO + 2H^+(aq) + 2e^-$$

$$HCHO + H_2O \rightarrow HCOOH + 2H^+(aq) + 2e^-$$

$$CH_3CH(OH)CH_3 \rightarrow CH_3COCH_3 + 2H^+ + 2e^-$$

These oxidations can be achieved by the use of acidified potassium dichromate or potassium permanganate. The yield in these reactions is often limited by the solubility of the oxidant. Yields can be increased by using alternative cations. For example, sodium dichromate may be used in alcohols instead of the potassium salt; tetrabutylammonium permanganate is more soluble in organic solvents than the potassium salts and so is a better oxidant for aldehydes. The latter compound is readily prepared from $KMnO_4$ and $(C_4H_9)_4NBr$; it would appear to be safer to prepare this compound fresh for each oxidation.

(e) Reduction reactions

Amines are prepared by the reduction of compounds containing carbon–nitrogen multiple bonds:

$$CH_3CH{=\!=}N \cdot OH + 2H_2 \rightarrow CH_3CH_2NH_2 + H_2O$$

$$CH_3 - C{\equiv}N + 2H_2 \rightarrow CH_3CH_2NH_2$$

Aniline (aminobenzene) can be obtained by the reduction of nitrobenzene:

$$2C_6H_5NO_2 + 6Sn + 14HCl(aq) \rightarrow [C_6H_5{}^+NH_3]_2 \, SnCl_4{}^{2-} + 4H_2O + 5SnCl_2(aq)$$

$$[C_6H_5{}^+NH_3]_2 \, SnCl_4{}^{2-} + 2NaOH(aq) \rightarrow 2C_6H_5NH_2 + 2H_2O + Na^+{}_2 SnCl_4{}^{2-}(aq)$$

(f) Aromatic reactions

The hydrogen atoms of aromatic systems, such as benzene, can be replaced by nitro groups ($-NO_2$), sulphonic acid groups ($-SO_3H$), halogens ($-Br$) alkyl or acyl groups ($-COR$):

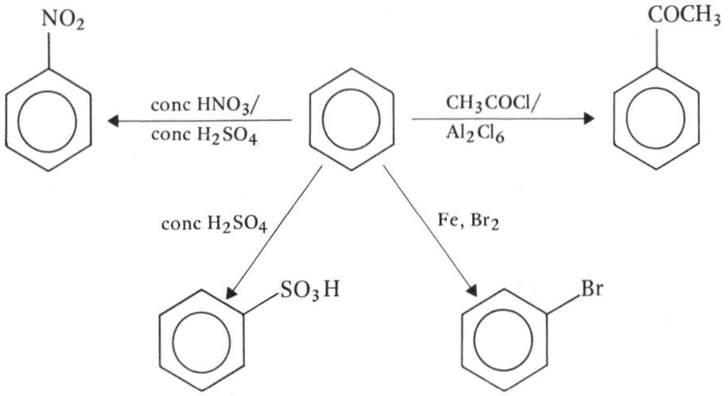

Homologous series

Organic compounds consist of carbon–hydrogen 'skeletons' on to which are attached functional groups. Compounds containing the same functional groups on carbon chains of differing lengths are said to belong to an homologous series (*Fundamentals of Chemistry*, p. 207). Because the functional part of the molecule is identical, the preparations and properties of each member of a series are very similar, if not identical. A few basic series are considered.

Halogenoalkanes

1. *Laboratory preparations*

The halogenoalkanes can be prepared in the laboratory by the following methods.

(a) Addition of hydrogen halides to alkenes   Reactive molecules such as hydrogen bromide attack the π-bond of an alkene:

When an asymmetric alkene is used, the nature of the product is governed by Markownikov's rule. In an addition reaction of a molecule of the form HX to an unsymmetrical alkene, the hydrogen atom is attached to the least substituted carbon atom (that is, the one with the largest number of hydrogen atoms bonded to it).

$$CH_3 - CH = CH_2 + HBr \rightarrow CH_3 - \underset{\underset{\text{Br}}{|}}{CH} - \underset{\underset{\text{H}}{|}}{CH_2}$$

the least substituted
carbon atom

As a result, this method tends to generate secondary or tertiary halogen compounds, that is, those in which the carbon atom of the C—X bond is attached to two or three other alkyl groups respectively.

(b) Halogenation of alcohols   Alcohols react with covalent chlorides to give halogenoalkanes. Hydrogen chloride is passed through a mixture of alcohol and anhydrous zinc chloride catalyst which is heated under reflux:

$$CH_3 CH_2 OH + HCl \rightarrow CH_3 CH_2 Cl + H_2 O$$

A similar reaction is possible for hydrogen bromide, though it is usually performed with potassium bromide and concentrated sulphuric acid. The salt and acid generate hydrogen bromide and the acid also acts as a catalyst.
    Alternatively the phosphorus halides may be used. This method is particularly suitable for the iodide, which is generated by using red phosphorus and iodine:

$$2P + 3I_2 \rightarrow 2PI_3$$

$$3CH_3 CH_2 OH + PI_3 \rightarrow 3CH_3 CH_2 I + H_3 PO_3$$

Phosphorus(V) chloride is normally used rather than the elements or phosphorus(III) chloride:

$$CH_3 CH_2 OH + PCl_5 \rightarrow CH_3 CH_2 Cl + POCl_3 + HCl$$

A very useful chlorinating agent is sulphinyl chloride (also known as thionyl chloride), $SOCl_2$. This has the advantage that the secondary products are gases, and the excess sulphinyl chloride is volatile and so is easily removed:

$$CH_3 CH_2 CH_2 OH + SOCl_2 \rightarrow CH_3 CH_2 CH_2 Cl + SO_2 + HCl$$

2. Industrial preparations

Industrially, chloromethane is obtained by the gas phase halogenation of methanol:

$$CH_3 OH + HCl \xrightarrow[\text{catalyst}]{\text{heat}} CH_3 Cl + H_2 O$$

This is used in the production of silicones, rubbers and other methylated compounds.

Chloroethane may be prepared from ethene as described above or by the chlorination of ethane:

$$CH_3 \cdot CH_3 + Cl_2 \xrightarrow{400\,^\circ C} CH_3 CH_2 Cl + HCl$$

The hydrogen chloride is then used in an addition reaction with ethene to produce more of the required halide. Chloroethane is used as a local anaesthetic and in the production of the antiknock tetraethyl-lead.

Bromoethane, manufactured from methanol and hydrogen bromide, is used as a fumigating agent. Bromoethane (from ethene and hydrogen bromide) is used in drug manufacture.

The fluoroalkanes are unreactive and are most commonly prepared by the displacement of chlorine atoms by fluorine using antimony(III) fluoride. For example, if trichloromethane is treated with antimony(III) fluoride, chlorodifluoromethane is formed and then decomposes at 700 °C to form tetrafluoroethane.

$$2CHCl_3 \xrightarrow{SbF_3} 2CHF_2 Cl \xrightarrow{700\,^\circ C} C_2 F_4 + 2HCl$$

This product is polymerised to give polytetrafluoroethane (PTFE, 'Teflon') which is chemically inert and so is used as a packaging material, even for very reactive chemicals.

3. Properties of the halogenoalkanes

The chemical properties of these derivatives can be divided into two categories.

(a) Elimination reactions The basic reaction is

$$-\underset{\underset{X}{|}}{\overset{\overset{H}{|}}{C}} - C - \longrightarrow \ \ \diagdown C = C \diagup \quad + HX$$

The elimination may be performed using a concentrated solution of alkali. However, the yields vary according to the halide type (see Table 7.1). It can be seen that elimination is unlikely with primary halides (see details of the substitution reactions below), but is the almost exclusive process for tertiary halides. Elimination is more likely than substitution if the base is in a hot,

Table 7.1 Yields in elimination reactions of halogenoalkanes, using alcoholic solutions of potassium hydroxide

Type	Example	Yield
Primary	CH_3CH_2Br	1% $CH_2{=}CH_2$
Secondary	$(CH_3)_2CHBr$	80% $CH_3{-}CH{=}CH_2$
Tertiary	$(CH_3)_3CBr$	100% $(CH_3)_2C{=}CH_2$

concentrated solution, but even under the conditions used in Table 7.1 it is the unfavourable reaction for the primary halides.

(b) Substitution reactions Primary halogenoalkanes usually react so as to replace the halogen atom by another atom or group. The iodo compounds are the most reactive; the fluorocompounds are the least reactive. Table 7.2

Table 7.2 Nucleophilic displacements of X (halogen) in RX

Nucleophile	Product
OH^-(aq)	ROH, alcohol
$OC_2H_5^-$ (in ethanol)	ROC_2H_5, ether
NH_2^- (in liquid ammonia)	RNH_2, amine
$CH_3CO_2^-$ (aq)	CH_3COOR, ester
CN^-(aq)	RCN, nitrile
SH^-(aq)	RSH, thiol
NH_3 (in ethanol, under pressure)	RNH_2, R_2NH, R_3N, amines as salts

lists some of the common nucleophiles (negatively charged reagents attacking positively polarised carbon atoms) and their products. Other nucleophilic substitutions are known, but these are the most common examples. In view of the risk of elimination reactions occurring (as described above), it is best to use the weakest base conditions practicable. For example, reaction with alkali is more likely to generate alcohols if silver oxide in moist ethoxyethane ('ether') is used.

Other reactions of the halogenoalkanes include:

(c) Friedel–Craft's alkylation Halogenoalkanes react with aromatic hydrocarbons in the presence of anhydrous aluminium chloride:

CH_3Cl + ⬡ $\xrightarrow{Al_2Cl_6}$ ⬡$-CH_3$

methylbenzene (toluene)

(d) Reduction Halogenoalkanes are reduced by catalytic hydrogenation (using nickel, platinum or palladium) or by active metals with a solvent

(e.g. Zn/HCl(aq), Na/CH$_3$CH$_2$OH). Lithium tetrahydroaluminate (LiAlH$_4$) in dry ethoxyethane is also an effective reducing agent. The corresponding alkanes are produced in each case.

$$CH_3CH_2Br + H_2 \rightarrow CH_3CH_3 + HBr$$

(e) Metallation   Lithium reacts with ethereal solutions of the halogeno-alkanes to give a metallated product:

$$CH_3CH_2CH_2CH_2Br + 2Li \rightarrow CH_3CH_2CH_2CH_2Li + LiBr$$

This compound, butyl-lithium, is a very reactive species which adds to multiple bonds

$$RLi + CH_2{=}CH_2 \rightarrow \underset{\underset{R}{|}}{CH_2}{-}CH_2{-}Li$$

and metallates aromatic compounds in the presence of a base:

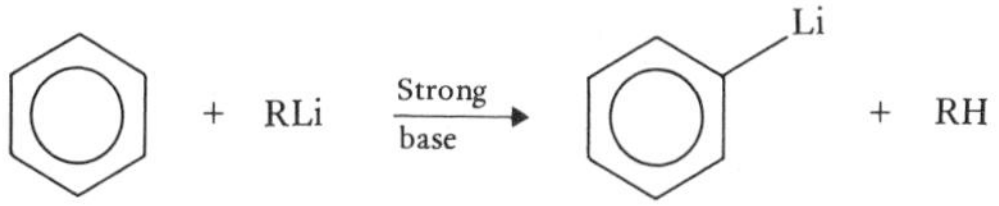

Sodium produces an alkyl–metal compound which reacts with excess halogenoalkane to give an alkane (the 'Wurtz reaction'):

$$CH_3CH_2Br + 2Na \rightarrow CH_3CH_2Na + NaBr$$

$$CH_3CH_2Na + CH_3CH_2Br \rightarrow CH_3CH_2CH_2CH_3 + NaBr$$

Probably the best known, and synthetically useful, reaction is the Grignard reaction. When a halogenoalkane, or an aromatic halide, is reacted with magnesium, an alkyl (or aryl) magnesium halide is formed.

$$RX + Mg \rightarrow RMgX$$

The aromatic halides are much less reactive than the halogenoalkanes and the reaction has to be catalysed with iodine or a trace of iodomethane.

The reaction is normally performed in an ethereal solvent, ethoxyethane (CH$_3$CH$_2$OCH$_2$CH$_3$) or tetrahydrofuran:

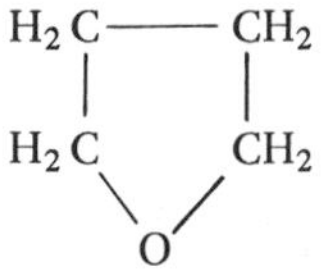

Figure 7.1 shows the structure of the 'Grignard reagent' as prepared from bromoethane in ethoxyethane. The reagent exists in several forms in equilibrium with each other:

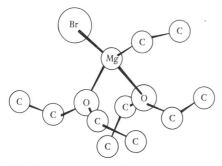

Fig. 7.1 Structure of ethylmagnesium bromide (2 molecules of ethoxy-ethane solvation; hydrogens not shown)

$$R-Mg \overset{X}{\underset{X}{\diagdown\diagup}} Mg-R \;\rightleftharpoons\; 2RMgX \;\rightleftharpoons\; RMg^+ + RMgX_2^- \;\rightleftharpoons\; R_2Mg + MgX_2$$

$$\quad\quad\quad II \quad\quad\quad\quad\quad I \quad\quad\quad\quad\quad III \quad\quad\quad\quad\quad IV$$

This equilibrium relationship is known as the **Schlenk equilibrium** after Schlenk and Schlenk who first postulated it. The solvent molecules have been omitted for simplicity, but all the species are solvated. The basic form, I, is that shown in Fig. 7.1. It dimerises to give the associated form, II, and undergoes ionisation into form III. The presence of the ionic species is indicated by the electrical conductivity of the solution; the presence of magnesium in both ions is apparent from the fact that it migrates to both electrodes. The reagent can also be produced from compounds IV.

The Grignard reagent has a number of synthetic applications which may be classified by its reaction types.

(i) With replaceable hydrogen atoms   If any water is present (e.g. in the solvent), it reacts with the Grignard reagent. The hydrogen displaces the magnesium halide grouping to generate an **alkane**:

$$RMgX + H_2O \rightarrow RH + Mg(OH)X$$

A similar reaction occurs with dilute acids:

$$RMgX + HCl(aq) \rightarrow RH + MgXCl$$

Alkynes also react with Grignard reagents in a similar manner:

$$RMgX + R' \cdot C{\equiv}CH \rightarrow RH + R' \cdot C{\equiv}C \cdot MgX$$

The products include a new Grignard reagent, one containing an alkynic group.

(ii) With reactive π-bonds   A characteristic of π-bonded species is that they undergo addition reactions with polar compounds. The same applies when a Grignard reagent reacts with an unsaturated compound:

$$\begin{array}{c}\delta\text{-} \quad \delta\text{+} \\ R \cdot Mg\ X\end{array} + \quad {>}C{=}O \rightarrow \begin{array}{c}|\\ -C-OMgX\\ |\\ R\end{array}$$

The addition product is then treated with hydrochloric acid; a reaction of type (i) occurs:

$$\begin{array}{c}|\\ -C-OMgX\\ |\\ R\end{array} + HCl \rightarrow \begin{array}{c}|\\ -C-OH\\ |\\ R\end{array} + MgXCl$$

The product is an **alcohol**. However, it is an alcohol with an even longer carbon chain than that in the Grignard reagent. Table 7.3 shows how the choice of carbonyl compound determines the nature of the alcohol formed.

Table 7.3   Reaction of RMgX with carbonyl compounds

Carbonyl compound		Alcohol produced	
Type	Formula	Type	Formula
Methanol	HCHO	Primary	RCH_2OH
Aldehydes	$R'CHO$	Secondary	$RCH(R')OH$
Ketones	$R'COR''$	Tertiary	$RCR'(R'')OH$

Carbon dioxide also reacts with the reagent, giving a **carboxylic acid**:

$$RMgX + O{=}C{=}O \rightarrow R-C{\Large\langle}\begin{array}{c}O\\ OMgX\end{array} \overset{H+}{\rightarrow} R-C{\Large\langle}\begin{array}{c}O\\ OH\end{array}$$

An analogous reaction occurs with carbon disulphide

$$RMgX + CS_2 \overset{H^+}{\rightarrow} R-C{\Large\langle}\begin{array}{c}S\\ SH\end{array}$$

and with sulphur dioxide

$$RMgX + SO_2 \overset{H^+}{\rightarrow} R-S{\Large\langle}\begin{array}{c}O\\ OH\end{array}$$

The nitrile group reacts in a manner similar to the carbonyl to give an imine:

$$RMgX + R' \cdot C{\equiv}N \rightarrow \begin{array}{c}R'-C{=}N-MgX\\ |\\ R\end{array}$$

The imine, when treated with acid, is hydrolysed to a **ketone**; $R'-CO-R$. An **amine** can be formed if the imine intermediate is reduced with lithium tetrahydroaluminate; $R'-CH-NH_2$ is obtained.
$$\begin{array}{c}|\\ R\end{array}$$

The alkenic group is unreactive towards the reagent unless it is chemically activated. This may be achieved by the use of nickel(II) chloride catalyst:

$$RMgX \quad + \quad \underset{}{>}C{=}C\underset{}{<} \xrightarrow[NiCl_2]{} \quad -\underset{\underset{R}{|}}{C}-\underset{\underset{MgX}{|}}{C}-$$

or by the presence of an amine function in the molecule:

$$RMgX \quad + \quad R'CH{=}CHCH_2NH_2 \rightarrow R'CH\underset{\underset{XMg}{|}}{-}CHCH_2NH_2$$

(iii) With epoxides   Epoxides are small ring cyclic ethers, such as epoxyethane:

$$\overset{\displaystyle O}{\underset{CH_2 {-\!\!-\!\!-} CH_2}{\diagup\!\diagdown}}$$

The Grignard reagent opens the ring and generates a primary alcohol with two extra carbon atoms over those in the reagent:

$$RMgX \quad + \quad \overset{\displaystyle O}{\underset{CH_2{-\!\!-}CH_2}{\diagup\diagdown}} \rightarrow RCH_2CH_2OMgX \xrightarrow{H^+} RCH_2CH_2OH$$

Epoxypropane reacts similarly:

$$RMgX \quad + \quad \overset{\displaystyle O}{\underset{CH_3CH{-}CH_2}{\diagup\diagdown}} \xrightarrow{H^+} \underset{\underset{CH_3}{|}}{RCHCH_2OH}$$

(iv) With chlorides   Binary chlorides are useful routes to other **organometallic compounds,** that is compounds containing organic groups attached to metals (or closely related elements). Lead(II) chloride reacts with phenylmagnesium bromide to give tetraphenyl-lead:

$$4C_6H_5MgBr + 2PbCl_2 \rightarrow (C_6H_5)_4Pb + 4MgBrCl + Pb$$

Silicon(IV) chloride reacts with the reagent to give a partially substituted compound:

$$4C_6H_5MgBr + 2SiCl_4 \rightarrow 2(C_6H_5)_2SiCl_2 + 4MgBrCl$$

These chloro derivatives of silicon can be hydrolysed

$$(C_6H_5)_2SiCl_2 + 2H_2O \rightarrow (C_6H_5)_2Si(OH)_2$$

and then undergo a loss of water

$$2(C_6H_5)_2Si(OH)_2 \rightarrow (C_6H_5)_2Si{-}O{-}Si(C_6H_5)_2$$

to give compounds known as siloxanes. Less extensive substitution of the silicon(IV) chloride provides a route to the **silicones:**

$$n(C_6H_5)SiCl_3 \rightarrow nC_6H_5Si(OH)_3$$

$$\begin{array}{ccccc}
 & C_6H_5 & & C_6H_5 & \\
 & | & & | & | \\
-Si & -O- & Si & -O- & Si- \\
 & | & & | & | \\
 & O & & O & C_6H_5 \\
 & | & & | & C_6H_5 \\
 & & & & | \\
C_6H_5-Si & -O- & Si & -O- & Si- \\
 & | & & | & | \\
 & & & C_6H_5 & \\
\end{array}$$

(v) With d-block metal salts   The salts of d-block metals cause coupling of the organic groups in Grignard reagents:

$$2C_6H_5MgBr \xrightarrow{CoCl_2} C_6H_5 \cdot C_6H_5 + MgBr_2 + Mg$$

In order to avoid this as a side-reaction, the magnesium used to prepare the Grignard reagent must be pure.

This coupling reaction also occurs with other reactive halides, and so iodides (which are the most reactive of the halogenoalkanes) are not normally used in the preparation of the Grignard reagent. Aliphatic chlorides are suitable reagents, but aromatic chlorides are too unreactive and bromides are normally used.

Polyfunctional compounds

When an organic molecule contains several functional groups which are separated by a carbon chain, the properties of the molecule are the sum of the properties of the individual groups. For example, amino acids contain the amine and carboxylic acid functions and display the properties of both:

$$CH_3-\underset{\underset{NH_2}{|}}{CH}-COOH$$

Alternatively, the groups might be attached directly to each other; in this case, their activities are often modified. This can be illustrated by the carbonyl group.

1.   Acyl chlorides

The acyl group is of the form $\underset{R}{\overset{R}{\diagdown}}C=O$ as found in aldehydes, ketones and

carboxylic acids. So, the acyl chlorides have the formula $\underset{Cl}{\overset{R}{\diagdown}}C=O$

They are named as illustrated by these examples:

CH_3COCl	ethanoyl chloride
CH_3CH_2COCl	propanoyl chloride

No reactions of the carbonyl group are observed since the polarity of the carbonyl function is offset by the electronegativity of the chlorine atom (see p. 218). On the other hand, the activity of the chlorine is enhanced compared to its reactivity in the halogenoalkanes. The chlorine atom is displaced by several groups as shown in the following reactions.

(a) Hydrolysis Whereas the hydrolysis of halogenoalkanes is slow and requires an alkali, water is adequate for the acyl chlorides:

$$CH_3 COCl + H_2 O \rightarrow CH_3 COOH + HCl$$

They must, therefore, be stored in a moisture-free atmosphere.

(b) Alcoholysis Reaction with alcohols gives an ester:

$$CH_3 COCl + CH_3 CH_2 OH \rightarrow CH_3 CO \cdot OCH_2 CH_3 + H_2 O$$
$$\text{ethyl ethanoate}$$

The reaction is more efficient than that between an acid and an alcohol.

(c) Ammonolysis Amides can be prepared very readily by reaction with ammonia:

$$CH_3 COCl + NH_3 \rightarrow CH_3 CONH_2 + HCl$$
$$\text{ethanamide}$$

(d) Friedel–Craft's reaction In the presence of anhydrous aluminium chloride, acyl chlorides react with aromatic hydrocarbons to give ketones.

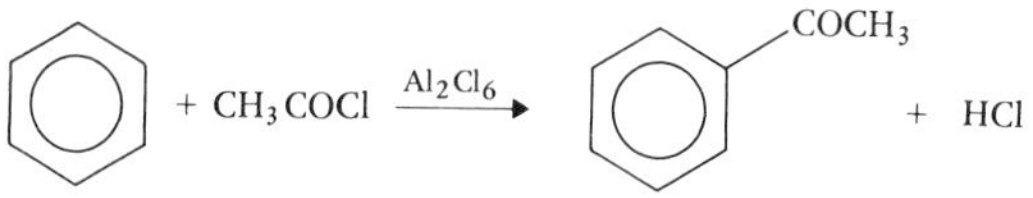

$$\bigcirc + CH_3 COCl \xrightarrow{Al_2Cl_6} \bigcirc^{COCH_3} + HCl$$

2. Amides

Whereas the carbonyl group activates a chlorine atom, it deactivates the amine group. Amides have the structure

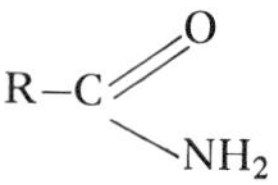

$$R-C {\overset{\displaystyle O}{\underset{\displaystyle NH_2}{}}}$$

The name is obtained by the addition of 'amide' to the alkane stem:

$CH_3 CONH_2$ ethanamide
$CH_3 CH_2 CONH_2$ propanamide

The carbonyl group is again inactive.

The properties of the amine group in amides may be compared with the properties of the same group in primary amines.

(a) Ethanamide is a weaker base than aminoethane. The pK_b values are

$CH_3 CH_2 NH_2$ 3.27
$CH_3 CONH_2$ 15.1

The decreased base strength is due to the electron-withdrawing effect of the carbonyl group reducing the electron density on the nitrogen. While it does form salts with acids

$$CH_3CONH_2 + HCl(aq) \rightarrow CH_3CO\overset{+}{N}H_3\overset{-}{Cl}(aq)$$

its high pK_b value means that it also reacts with bases to form salts:

$$2CH_3CONH_2 + HgO \rightarrow (CH_3CONH)_2Hg + H_2O$$

As a result of its poor basic character, it does not coordinate with metals (*Fundamentals of Chemistry*, p. 231).

(b) Nitrous acid reacts with amides to give the corresponding acid and nitrogen.

$$CH_3CONH_2 + HNO_2 \rightarrow CH_3COOH + N_2 + H_2O$$

(c) Amides can be hydrolysed to the corresponding acids, or the salts, when heated under reflux with acid, or base. Aqueous sodium hydroxide can be replaced with an alcoholic solution of potassium hydroxide for the less soluble amides.

$$CH_3CONH_2 \quad + \quad H_2O \quad\quad \rightarrow CH_3COONH_4$$
$$CH_3COONH_4 + HCl(aq) \quad \rightarrow CH_3COOH \quad + NH_4Cl$$
$$CH_3COONH_4 + NaOH(aq) \rightarrow CH_3COONa \quad + NH_3 + H_2O$$

This is a reaction which contrasts with the properties of amines.

(d) Amides may also be dehydrated by the use of P_4O_{10} or ethanoic anhydride; nitriles are formed.

$$6CH_3CONH_2 + P_4O_{10} \rightarrow 6CH_3CN + 4H_3PO_4$$

(e) Amines are formed in the reaction of amides with bromine and concentrated sodium hydroxide solution. The carbon chain length is decreased. This is known as the Hofmann degradation of amides.

$$CH_3CONH_2 + Br_2 + 4NaOH \rightarrow CH_3NH_2 + Na_2CO_3 + 2NaBr + 2H_2O$$

In summary, we see from these two examples of polyfunctional compounds that the properties of the functional groups are modified when they are adjacent; in one case, the chlorine atom is more reactive, and in the other case, the amino group is less basic than in the monosubstituted derivatives. In both cases the properties of the carbonyl group are extensively modified.

Polysubstitution of benzene

The substitution reactions of benzene have been described (p. 187) as producing monosubstituted derivatives. However, further substitution is also possible. For example, nitration of benzene at moderate temperatures (below $50\,^{\circ}C$) leads to nitrobenzene, but above that temperature further reaction occurs to give 1,3-dinitrobenzene:

Further substitution is only possible with higher temperatures and stronger reagents.

Sulphonation by concentrated sulphuric acid is a slow reaction which gives a relatively low yield of the sulphonic acid. Further substitution is very

difficult. The sulphonation of benzene proceeds more readily if chloro-sulphonic acid (Cl · SO_3H) is used instead of sulphuric acid.

The rate at which the introduction of a second substituent into benzene occurs is dependent on the nature of the first substituent. For example, we may compare the rate of nitration of several substituted benzenes (Table 7.4). Groups that enhance the rate of substitution relative to the hydrogen

Table 7.4 Relative rates of nitration

Group X in C_6H_5X	Relative rate of nitration
NH_2	10^6
CH_3	25
H	1
Cl	0.033
NO_2	10^{-6}

atom are described as **activating** groups; those causing a reduction in the rate of substitution are **deactivating**. So, in Table 7.4, the amino and methyl groups are activating; the chloro and nitro groups are deactivating. In fact, the aminobenzene (aniline) may be nitrated using a simple nitrate in acid medium. Concentrated acids are liable to generate a dinitro derivative of the aminobenzene.

A further example of this effect is in the bromination reaction. Benzene may be brominated using elemental bromine and an iron catalyst. The benzene ring of benzaldehyde, C_6H_5CHO, is not attacked by bromine. But phenol, C_6H_5OH, reacts with bromine so readily that it undergoes *tri*-bromination even with a dilute aqueous solution of bromine.

The positions of substitution vary. In the case of the nitration of nitro-benzene, the second substituent occupies the 3-position relative to the first group. Aminobenzene, on the other hand, gives a mixture of 1-amino-2-nitro- and 1-amino-4-nitrobenzenes. All the compounds fall into one or other of these two classes. Table 7.5 classifies the substituents according to whether they give 2-, 4-substituted compounds (class I) or 3-substituted compounds (class II).

Table 7.5   Classification of groups by directional effects

Class I	Class II
OH	$\overset{+}{N}H_3$
NR_2	NO_2
NHR	SO_3H
NH_2	CHO
OR	$COCH_3$
Cl, Br, I	CN
R	COOH
	COOR

The class II compounds involve π-bonded electronegative oxygen atoms

or positively charged atoms

all of which exert a strongly electron-withdrawing effect on the ring (see p. 219). With the exception of the halogens, all the class I groups have an activating effect on the aromatic nucleus. The halogens and class II are deactivating groups.

Alternative names have been used for these disubstituted compounds, in which the prefix indicates the relative positions of the substituents:

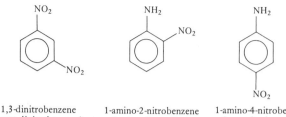

| 1,3-dinitrobenzene (meta-dinitrobenzene) | 1-amino-2-nitrobenzene (ortho-nitroaniline) | 1-amino-4-nitrobenzene (para-nitrobenzene) |

2- and 4- substitution normally occur in the same reaction, the directing effects (see p. 199) being identical. Theoretically, the proportions of the products should be in a 2 : 1 ratio respectively, since there are two similar positions in the first case (nominally 2- and 6-), but only one in the second case. For example, nitration of methylbenzene gives 61 per cent of the 2-nitro compound and 39 per cent of the 4-nitro derivative. In many cases the proportion of the 4-substituted compound predominates due to steric hindrance; that is, a bulky group hinders substitution on the neighbouring carbon atoms. For example, when methylbenzene (toluene) is nitrated, the 2- and 4- nitrotoluenes are produced in an approximately 2 : 1 ratio. However, chlorobenzene gives *only* the 4-substituted sulphonic acid when treated with concentrated sulphuric acid. The reason for the absence of any 2-chlorosulphonic acid would seem to be due to the size of the two groups. Figure 7.2 illustrates the relative sizes of the groups in these two cases.

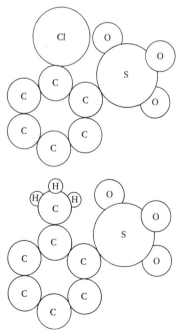

Fig. 7.2   Steric hindrance in chlorobenzene and methylbenzene

The extent to which a benzene derivative can undergo further substitution can be controlled through the choice of the temperature, concentration of reagents, choice of solvent and by adaptation of the substrate. An example of the manner in which the substrate can be modified in order to vary the reaction products would be in the preparation of the diaminobenzenes (Fig. 7.3).

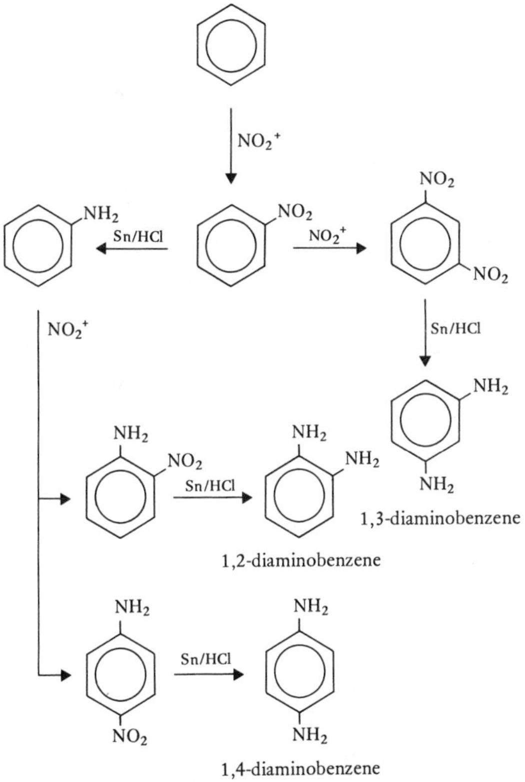

Fig. 7.3   Preparation of diaminobenzenes

Preparation of 'high polymers'

The term **high polymer** is applied to molecules containing large numbers (hundreds or thousands) of interlinked simple units called monomers. Consider the following high polymers, the repeating unit being framed for easy identification.

Polyethene ('polythene'):

$$-CH_2-CH_2-\boxed{CH_2-CH_2}-CH_2-CH_2-$$

(the grouping $CH_2 —CH_2$ is used as the basic entity here because it is derived from ethene).

Polychloroethene (polyvinylchloride, PVC):

$$—CH—CH_2 +CH—CH_2 +CH—CH_2 —$$
$$\quad\ \ |\qquad\qquad\ \ |\qquad\qquad\ \ |$$
$$\quad\ \ Cl\qquad\qquad\ Cl\qquad\qquad\ Cl$$

Polyacrylonitrile:

$$—CH—CH_2 +CH—CH_2 +CH—CH_2 —$$
$$\quad\ \ |\qquad\qquad\ \ |\qquad\qquad\ \ |$$
$$\quad\ \ CN\qquad\qquad CN\qquad\qquad CN$$

Polyphenylethene (polystyrene):

$$—CH—CH_2 +CH—CH_2 +CH—CH_2 —$$
$$\quad\ \ |\qquad\qquad\ \ |\qquad\qquad\ \ |$$
$$\quad\ \ C_6H_5\qquad\quad C_6H_5\qquad\quad C_6H_5$$

Poly(2-chloro-1,3-butadiene) (neoprene rubber):

$$—CH_2 —CH{=}C—CH_2 +CH_2 —CH{=}C—CH_2 +CH_2 —CH{=}C—CH_2 —$$
$$\qquad\qquad\qquad |\qquad\qquad\qquad\qquad\qquad |\qquad\qquad\qquad\qquad\qquad |$$
$$\qquad\qquad\qquad Cl\qquad\qquad\qquad\qquad\qquad Cl\qquad\qquad\qquad\qquad\qquad Cl$$

Polymethylmethacrylate ('Perspex')

$$\qquad\quad CH_3\qquad\quad CH_3\qquad\quad CH_3$$
$$\qquad\quad\ |\qquad\qquad\ |\qquad\qquad\ |$$
$$—CH_2 —C—CH_2 —C—CH_2 —C—$$
$$\qquad\quad\ |\qquad\qquad\ |\qquad\qquad\ |$$
$$\qquad\ {}_C OOCH_3\quad {}_C OOCH_3\quad {}_C OOCH_3$$

Polytetrafluoroethene ('PTFE', 'Teflon'):

$$—CF_2 —CF_2 +CF_2 —CF_2 +CF_2 —CF_2 —$$

Nylon-6:

$$—CO(CH_2)_5 —NH +CO(CH_2)_5 —NH +CO(CH_2)_5 —NH—$$

(the number 6 in the name indicates that there are 6 carbon atoms in the repeating unit).

Poly(ethylterephthalate) ('Terylene'):

Phenol—methanal resins (e.g. 'Bakelite'):

Many other high polymers are known, but these examples are sufficient to illustrate the variety encountered.

The polymers fall into two categories according to their methods of preparation.

(a) Addition polymers

If an unsaturated compound, such as ethene, is polymerised, the larger unit is formed by the monomer units adding to each other by cleavage of the π-bond:

$$CH_2\!=\!CH_2 + CH_2\!=\!CH_2 \rightarrow \cdot CH_2\!-\!CH_2\!-\!CH_2\!-\!CH_2 \cdot$$

The terminal units have unpaired electrons which can attack other alkenic molecules or abstract atoms from other compounds.

The initiation of this process (the breaking of at least one π-bond) is achieved by the use of ultraviolet light, by the use of a peroxide such as 'benzoylperoxide', or by heat and pressure. An American process uses butyllithium (p. 191) for this initiation step.

$$CH_2\!=\!CH_2 \overset{h\nu}{\rightarrow} \cdot CH_2\!-\!CH_2 \cdot$$

(b) Condensation polymers

Other polymers are obtained by condensation reactions. Reaction of 1,6-diaminohexane and hexan-1,6-dioic acid ('adipic acid'), especially as the acyl chloride generates nylon-6,6:

$$n\,[H_2N\,(CH_2)_6\,NH_2 + HOOC\,(CH_2)_4\,COOH] \rightarrow$$
$$[NH\,(CH_2)_6\,NH\!-\!CO\,(CH_2)_4\,CO]_n + nH_2O$$

Bakelite involves a condensation reaction between the 2- and 4- hydrogens of phenol and the carbonyl group of methanal:

Nylon-6 is obtained by an alternative method. Cyclohexanone is converted to its oxime by a condensation reaction with hydroxyammonia (see p. 186):

This oxime is then rearranged to 'caprolactam', which is polymerised to nylon-6.

Biological chemicals often exist as polymers. For example, proteins are polyamino acids:

$$n \quad \underset{\underset{NH_2}{|}}{\overset{\overset{R}{|}}{CH}}—COOH \qquad (NH—\underset{|}{\overset{\overset{R}{|}}{CH}}—CO)_n$$

In proteins, the group R varies along the chain. There are twenty commonly occurring 2-amino acids (also known as α-amino acids). The sequence of these acids, and the frequency with which each occurs in the chain, varies from protein to protein. Because of the large number of permutations of these twenty units in a protein, a large number of structures are possible. Proteins are involved in the many enzymes of living systems and in such structural features as muscles, hair, scales, collagen and silk. They occur in egg albumin, haemoglobin, casein and in the polio virus. However, only a tiny percentage of the possible protein compositions lead to biologically useful molecules.

Carbohydrates such as starch, cellulose and glycogen are polysaccharides. Cellulose is a condensation polymer of the monosaccharide β-glucose:

Organic synthesis

The organic chemist is often required to prepare a new substance from readily available starting materials. While a number of routes may be possible 'on paper', often several of these are impracticable. They may involve steps with poor yields, dangerous reactions, volatile compounds which are difficult to handle, and so on. The choice of reagents, the order of a series of substitutions, etc. all have to be considered carefully. The number of steps is significant in terms of the yield of material (if not for the time factor). For example, if we are starting with 1 g of reactant and the yield at each step is 60 per cent only, then consider the overall yield in routes involving two and five steps respectively:

(i) 2 steps:

$$A \rightarrow B \longrightarrow Product$$
$$1\ g \quad 0.6\ g \quad 0.36\ g$$

(ii) 5 steps:

$$A \rightarrow B \longrightarrow C \longrightarrow D \longrightarrow E \longrightarrow Product$$
$$1\ g \quad 0.6\ g \quad 0.36\ g \quad 0.22\ g \quad 0.13\ g \quad 0.08\ g$$

This section considers the principles on which preparations are based by means of a number of examples.

1. Synthesis of aminoethane from ethanol

The first step is to consider:

(*a*) whether there is a change in the number of carbon atoms in the molecule; and

(*b*) which other atoms have been replaced by different atoms or groups.

In this synthesis the number of carbon atoms is unchanged. We must, therefore, avoid reactions involving the use of, say, cyanide (—CN) or the addition of Grignard reagents to multiple bonds ($\gtrdot C{=}O$), since these necessitate the introduction of extra carbon atoms (p. 193).

An oxygen atom is replaced by a nitrogen atom:

$$C{-}C{-}O \rightarrow C{-}C{-}N$$

This is achieved in a condensation reaction (p. 186):

$$C—C=O + H_2N—Y \rightarrow C—C=N—Y$$

The problem can now be considered in the steps

$$C—C—O \qquad\qquad C—C—N$$
$$\downarrow 1 \qquad\quad \xrightarrow{\ 2\ } \qquad \uparrow 3$$
$$C—C=O \quad H_2N \cdot Y \quad C—C=N—Y$$

Step 1 is an oxidation step, using acidified dichromate. Reduction by $LiAlH_4$ of an oxime is a convenient means of performing step 3 (p. 187). The group Y in step 2, therefore, should be the hydroxyl group and the reagent is hydroxy-ammonia. The synthesis may be summarised as:

$$CH_3CH_2OH \xrightarrow{K_2Cr_2O_7/H^+(aq)} CH_3CHO \xrightarrow{H_2NOH} CH_3CH=NOH$$

ethanol   ethanal   ethanal oxime

$$\xrightarrow{LiAlH_4} CH_3CH_2NH_2$$

aminoethane

2. *Synthesis of 2,2-dimethylethanol*

$$CH_3—\overset{\overset{\displaystyle CH_3}{|}}{\underset{\underset{\displaystyle CH_3}{|}}{C}}—OH$$

Three synthetic routes are considered: (*a*) the hydrolysis of the corresponding halide $(CH_3)_3C \cdot Br$, might be expected to yield the alcohol, but this leads to an elimination reaction with tertiary halides (p. 190); (*b*) the hydration of an alkene via the addition of concentrated sulphuric acid is more likely to be successful. The difficulty in this reaction is that the alkene, 2-methylpropene, is not likely to be readily available in the laboratory. Also,

$$CH_3—\underset{\underset{\displaystyle CH_3}{|}}{C}=CH_2 + H_2O \rightarrow CH_3—\overset{\overset{\displaystyle OH}{|}}{\underset{\underset{\displaystyle CH_3}{|}}{C}}—CH_3$$

being a gas, it is not convenient to handle; (*c*) an alternative route to tertiary alcohols is by the Grignard reaction with a ketone (p. 193).

$$CH_3MgBr + \overset{CH_3}{\underset{CH_3}{>}}C=O \rightarrow \overset{CH_3}{\underset{CH_3}{>}}C\overset{OH}{\underset{CH_3}{<}}$$

The reagent is prepared from a halogenomethane. As previously indicated, the chloro- and bromoalkanes are to be preferred to the iodoalkanes (p. 195).

In this case, it is more convenient to use iodoethane, even at the expense of the maximum yield because of the handling difficulties of the other two halides. The boiling points of these derivatives are:

CH_3Cl 249 K
CH_3Br 277 K
CH_3I 316 K

The 2,2-dimethylethanol can be prepared, then, by the preparation of the methylmagnesium iodide from iodoethane; the reaction of this with propanone, followed by treatment with acid, will give the required compound.

3. Synthesis of CH_3CH_2COOH from CH_3CH_2OH

Examination of these formulae indicates an increase in the number of carbon atoms by one. We must, therefore, consider the introduction of, say, a cyanide group or the use of a Grignard reagent. In the former case an acid is obtained by hydrolysis

$$RCN \xrightarrow{H^+/H_2O} RCOOH + NH_4^+$$

and the Grignard reaction requires carbon dioxide

$$RMgBr \xrightarrow[HCl(aq)]{CO_2} RCOOH + MgBrCl$$

Both of these substances are obtained from the same halogeno compound:

$$RCN \xrightarrow{CN^-} RX \xrightarrow{Mg} RMgX$$

Several routes are available for the preparation of the halides from alcohols (p. 188).

While the cyanide route involves the use of an extremely toxic material, potassium cyanide, the Grignard reaction requires the use of very dry, very pure reagents and apparatus. On balance the latter route is preferable since the reagents are generally readily available in the required state and the yield is better than that for the cyanide route.

$$CH_3CH_2OH \xrightarrow[H_2SO_4]{KBr} CH_3CH_2Br \xrightarrow{Mg} CH_3CH_2MgBr \xrightarrow[H^+(aq)]{CO_2} CH_3CH_2COOH$$

4. Synthesis of 3-nitrophenol from benzene

One of the most effective means of introducing a substituent into benzene is nitration.

But how do we introduce the phenolic grouping? The hydroxyl group cannot be introduced directly on to a benzene ring. It is obtained by the reaction of an amino group with nitrous acid (p. 186).

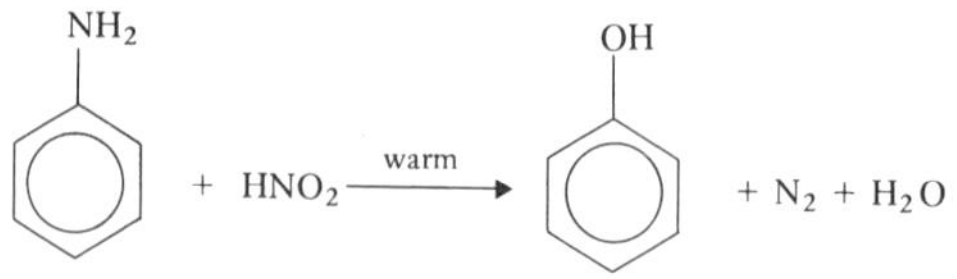

The amino group is obtained by the reduction of the nitro group (p. 187). However, phenol on nitration gives 2- and 4-nitrophenol, not 3-nitrophenol (p. 187). The 3-nitro grouping is obtained by the nitration of nitrobenzene. So, the route employed would involve this step, followed by the careful reduction of the product (control of the amount of reducing agent and time of reduction) to give 1-amino-3-nitrobenzene.

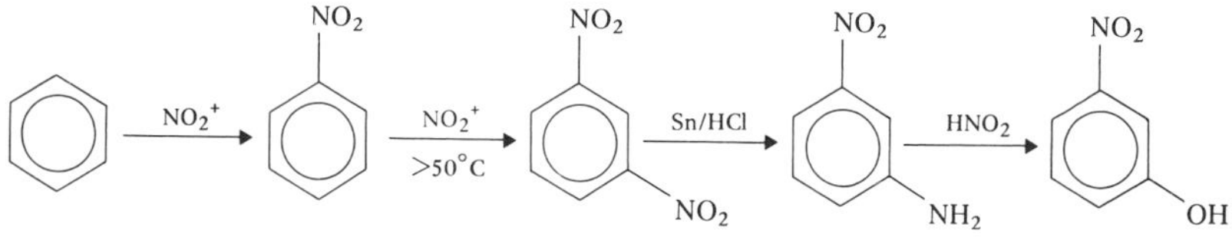

5. *Synthesis of 2-oxo-ethanoic acid (pyruvic acid) from ethanoic acid*

A comparison of the formulae

$$CH_3 —CH_2 —COOH \text{ and } CH_3 —CO—COOH$$

shows that the extra functional group has been introduced on the carbon atom adjacent to the carboxyl group. This carbon atom of an acid is readily attacked by chlorine:

$$CH_3 —CH_2 —COOH \xrightarrow{Cl_2} CH_3 —CH—COOH$$
$$\underset{Cl}{|}$$

The carbonyl group is produced by the oxidation of an alcohol. So, our synthetic pathway becomes:

$$CH_3 CH_2 COOH \qquad\qquad CH_3 COCOOH$$

A, C

$$CH_3 CHCl \cdot COOH - - - \xrightarrow{B} CH_3 CH(OH)COOH$$

The remaining step B is achieved by hydrolysis.

$$CH_3 CHClCOOH \xrightarrow{H_2O} CH_3 CH(OH)COOH$$

Table 7.6 lists some of the useful reagents for the introduction of functional groups.

Purification

When the synthesis is complete, it is necessary to isolate the product from unchanged reactants, other products and the solvent. While the detailed separation procedures vary with each specific case, a few general techniques may be described.

Table 7.6 Table showing the reagent to be used to convert a functional group in a substrate to a required functional group

Group required	Group in substrate	Reagent
(a) Aliphatic compounds		
$-CH_2R'$	$-CO \cdot R'$	Zn and HCl(aq)
$>C{=}C<$	$>CH \cdot CHX$	NaOH(alc)
$-Cl$	$-OH$	$SOCl_2$
$-Br$	$-OH$	KBr and H_2SO_4
	$>C{=}C<$	HBr
$-I$	$-OH$	P and I_2
$-OH$	$-X$	NaOH(aq)
	$-MgX$	$>C{=}O<$
	$>C{=}C<$	H_2SO_4
$-CN$	$-X$	KCN
$-CHO$	$-CH_2OH$	$K_2Cr_2O_7$(aq)
$-COR'$	$-CH(OH)R'$	$K_2Cr_2O_7$(aq)
$-CH_2NH_2$	$-CN$	$LiAlH_4$
	$-C{=}N \cdot OH$	$LiAlH_4$
$-COOH$	$-MgX$	CO_2
$-COOR'$	$-COOH$	$R' \cdot OH$
$-OR'$	$-X$	OR'^-
(b) Aromatic compounds		
$-R$	$-H$	RX and Al_2Cl_6
$-COR$	$-H$	RCOCl and Al_2Cl_6
$-X$	$-H$	X_2 and Fe
$-NO_2$	$-H$	HNO_3 and H_2SO_4
$-SO_3H$	$-H$	H_2SO_4
$-NH_2$	$-NO_2$	Sn and HCl(aq)
$-OH$	$-NH_2$	HNO_2

If the product, or impurity, is soluble in a solvent (e.g. ethoxyethane or water) in which the other substances are insoluble, and the solvent is immiscible with the original mixture, then it may be separated by solvent extraction (p. 103).

An illustration of this technique can be taken from the preparation of bromoethane from ethanol. The product is contaminated with hydrogen bromide (from concentrated sulphuric acid and potassium bromide) and sulphur dioxide (from the reduction of sulphuric acid by hydrogen bromide).

The bromoethane is washed with sodium carbonate solution to remove the acidic substances which have dissolved in the organic compound. Carbon dioxide is released as the acid is removed and neutralised. The organic layer is washed with water to remove the carbonate. Finally it is dried before redistillation.

Extraction of a base from an organic solvent into aqueous medium is facilitated by use of an aqueous mineral acid. For example, an amine may be prepared by the reduction of a nitrile (e.g. butanenitrile):

$$RCN + 4Na + 4C_2H_5OH \rightarrow RCH_2NH_2 + 4C_2H_5ONa$$

the amine can be separated from the mixture by treatment with hydrochloric acid. A salt is formed which dissolves in the aqueous solution. The aqueous layer is separated and the amine is recovered by the addition of base:

$$RCH_2NH_2 + HCl(aq) \rightarrow RCH_2\overset{+}{N}H_3\ Cl^-(aq)$$

$$RCH_2\overset{+}{N}H_3\ Cl^-(aq) + Na^+OH^-(aq) \rightarrow RCH_2NH_2 + Na^+Cl^-(aq) + H_2O$$

The amine can be isolated from the aqueous mixture by extraction into ether. The ether is removed by evaporation.

An acid may be separated by an analogous process. For example, benzoic acid is prepared by the Grignard reaction (p. 191):

$$C_6H_5MgBr + CO_2 + HCl \rightarrow C_6H_5COOH + MgBrCl$$

The acid is removed from the reaction mixture by extraction with sodium hydroxide solution

$$C_6H_5COOH + Na^+OH^-(aq) \rightarrow C_6H_5COO^-Na^+(aq) + H_2O$$

Neutralisation of this solution with acid and extraction with ethoxyethane isolates the benzoic acid.

A carbonyl compound can be extracted by the addition of sodium hydrogen sulphite:

$$CH_3CHO + Na^+HSO_3{}^-(aq) \rightarrow \begin{matrix} CH_3 \\ \diagdown \\ C \\ \diagup \diagdown \\ H \end{matrix} \begin{matrix} O^-Na^+ \\ \diagup \\ \\ \diagdown \\ SO_3H \end{matrix}$$

The sulphite adduct crystallises out and can be decomposed to regenerate the carbonyl compound by the addition of dilute acid.

Before an organic solvent is removed by evaporation or an organic compound is redistilled, any traces of moisture are removed. Various drying agents are available. Anhydrous calcium chloride will dry many solvents but it is not suitable for alcohols or amines, since these react with calcium chloride. Dried magnesium sulphate is a suitable drying agent for a wide range of solvents. If it is desired to remove the final traces of water (not usually necessary for the above situations) then molecular sieves are very effective. These are solid particles with small pores which are able to trap small molecules such as water. Table 7.7 lists some common desiccants for solvents and indicates substances that are likely to react with the drying agent.

Table 7.7 A selection of desiccants for organic solvents. The desiccants must not be used to dry solutions containing the substances listed.

Desiccant	Solvent	Reactants
Aluminium oxide	Hydrocarbons	
Calcium chloride	Ethers; esters; alkyl and aryl halides; hydrocarbons	Alcohols; amines; carbonyl compounds
Calcium oxide	Low relative molecular mass alcohols; amines	Acidic compounds; esters
Magnesium sulphate, anhydrous	Most compounds	
Molecular sieves	Limited by pore size only — see manufacturer's specifications	Small molecules
Phosphorus(V) oxide	Hydrocarbons; ethers; alkyl/aryl halides; nitriles	Alcohols; acids; amines; ketones
Potassium carbonate, anhydrous	Alcohols; nitriles; ketones; esters; amines	Acids; phenols
Potassium (or sodium) hydroxide	Amines	Acids; phenols; esters; amides
Silica gel	Most compounds	

The final stage of purification involves redistillation (of liquids) or recrystallisation (of solids). These processes are described in more detail in Chapter 4.

Summary

At the conclusion of this chapter, you should be able to:

1. give typical examples of addition, substitution, condensation, oxidation, reduction and aromatic substitution reactions;
2. describe, in outline, the laboratory preparation of a halogenoalkane;
3. describe elimination, substitution, alkylation, reduction and metallation reactions of the halogenoalkanes;
4. describe the nature, preparation and uses of the Grignard reagent;
5. describe the properties of the acyl chlorides and compare their reactivity with chloroalkanes;
6. describe the properties of the amides and compare their reactivity with the aminoalkanes;
7. distinguish between activating and deactivating groups in aromatic substitution reactions;
8. give systematic names for polysubstituted compounds;
9. distinguish between addition and condensation polymers;

212

10. determine the shortest practicable route from one organic compound to another of similar structure;
11. describe the main procedures for the purification of organic products.

Experiments

7.1 Preparation of bromoethane

Set up the apparatus as in Fig. 7.4. In the flask place ethanol (5 cm^3) and

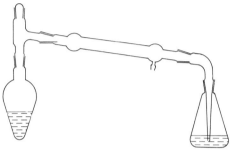

Fig. 7.4 Apparatus for the preparation of bromoethane

water (2.5 cm^3). Then add slowly, with shaking and cooling, concentrated sulphuric acid (6 cm^3). Add sodium bromide (5 g) in small quantities. Connect the flask to the condenser, add anti-bumping granules and heat the flask on a sand tray. The bromoethane will distil over and collect below the water in the receiver. The water minimises the loss of the volatile product by evaporation.

Collect the lower oily layer in a separating funnel and wash the oil with an equal volume of 2 M Na$_2$CO$_3$ solution (do not forget to release any carbon dioxide evolved). Remove the aqueous layer and wash the oil with water. Dry the oil over lumps of calcium chloride in a stoppered flask for 30 minutes. Redistil the product from dry apparatus collecting the 38–40 °C fraction. Determine the yield obtained.

7.2 Reactions of bromoethane and ethanoyl chloride

Place 5 per cent silver nitrate solution (1 cm^3) in a small test tube and add bromoethane (2 drops). Allow it to stand for 5 minutes and note any changes.

Repeat the reaction but add ethanoyl chloride carefully instead of the bromoethane.

Place some ethanol (1 cm^3) in a test tube and add bromoethane (2–3 drops). Repeat the process using ethanoyl chloride instead of the bromo-ethane. When the fumes of hydrogen chloride, if any, have cleared in each case, compare the odour of the products.

7.3 Preparation and application of a Grignard reagent

Use the apparatus shown in Fig. 7.5 with the gas inlet closed. All the

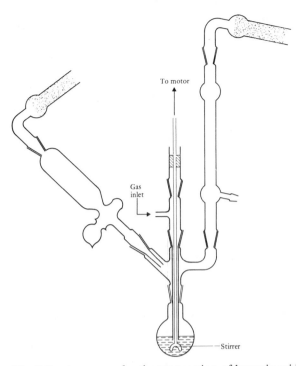

Fig. 7.5 Apparatus for the preparation of benzoic acid

apparatus and chemicals must be free of water. The liquid chemicals should be dried overnight over magnesium sulphate; the apparatus and magnesium turnings (high purity) must be dried in an oven.

Place magnesium (2.5 g) and ethoxyethane (25 cm³) in the flask; drop a crystal of iodine on to the metal. Put bromobenzene (15.7 g) in ethoxyethane (25 cm³) in a dropping funnel. Allow 5–10 cm³ to run in. The solution should fade and become opaque. If the reaction becomes too vigorous, cool it slightly so as to prevent side-reactions. If no reaction has occurred in 2 minutes, warm it in a water bath (no flames!). When the reaction begins, start the stirrer and allow the rest of the reagent to run in so as to maintain a gentle reflux. When the addition is complete, continue to reflux the mixture on a warm water bath for 30 minutes. The flask contains an ethoxyethane solution of the Grignard reagent.

Attach the gas inlet to a supply of dry CO_2. Cool the reagent in ice and pass a slow stream of CO_2 for 15 minutes. Decompose the carbonated reagent with 3 M hydrochloric acid. Remove the organic layer. Wash the aqueous layer with more ethoxyethane and combine the organic solutions. Add excess dilute ammonia solution to the organic solution, shake and separate the layers. (The organic layer contains diphenyl formed by a side-reaction.) Acidify the aqueous layer with dilute hydrochloric acid and filter off the benzoic acid. Recrystallise from hot water and determine its melting point.

7.4 Preparation of ethanamide

Add ammonium carbonate (4 g) to a pear-shaped flask and slowly add glacial ethanoic acid (10 cm^3). Allow the reaction to subside after each addition before more acid is added. When the addition is complete, boil under reflux, using an air condenser, for 30 minutes. Raise the temperature slowly until the temperature at the top of the condenser (still in the reflux position) is 200 °C. Change the receiver (acid and water are distilled off up to this point) and raise the temperature up to 225 °C. The amide will distil off between 200 and 225 °C. Recrystallise the solid product from propanone (acetone) and determine its melting point.

7.5 Properties of amides and amines

1. To ethanamide (0.1 g) in a test tube, add 2 M sodium hydroxide solution (2 cm^3). Boil gently and identify the gas evolved.
2. Add ethanamide (0.1 g) to water and test the product with pH paper. Repeat the test using aminoethane instead of the amide.
3. To a little of the amide in a test tube, add dilute hydrochloric acid. Repeat for the amine.
4. Dissolve the amide (0.1 g) in water (1 cm^3). Add sodium nitrite (0.2 g) and shake until it completely dissolves. Add sufficient dilute hydrochloric acid to just acidify the mixture. Repeat the test for the amine.

7.6 Examples of polymerisation

Condensation polymerisation

This reaction must be performed in a fume cupboard; use rubber gloves and eye protection.

In a small beaker, mix 40 per cent methanal solution (1 cm^3), phenol (3 g) and 40 per cent sodium hydroxide solution (0.5 cm^3). Heat the mixture gently.

Addition polymerisation

In a boiling tube, mix phenylethene ('styrene', 5 cm^3) and 2 M sodium hydroxide (5 cm^3). Shake for 1 minute, separate off the phenylethene and transfer it to another tube. Add dodecanoyl peroxide (0.1 g) and heat in a boiling water bath until the polymer sets.

References

Chemistry of the Grignard reagent; J. H. J. Peet, *Educ. Chem.*, 1968, **5**, 199–202.

Victor Grignard (a translation of his original paper); P. R. Jones and E. Southwick; *J. Chem. Ed.,* 1970, **47**, 290–9.

Constitution of the Grignard reagents; N. A. Bell, *Educ. Chem.,* 1973, **10**, 143–5.

Grignard reagents – composition and mechanisms of reaction; E. C. Ashby, *Quart. Rev.,* 1967, **21**, 259–85.

Organomagnesium compounds in organic synthesis; B. J. Wakefield, *Chem. Ind.,* 1972, 450–3.

An Introduction to the Design of Organic Synthesis; S. Turner (Koch-Light).

Experiments in Applied Chemistry; P. Tooley (J. Murray, 1975).

Polymer Chemistry Experiments (BDH).

Experiments in Polymer Chemistry; H. S. Finlay (Shell).

Polymers and Plastics in School Science Laboratories (ICI, 1973).

Polymers; R. W. Thomas (Pergamon, 1969).

Films

Synthesis of an organic compound (CHEM Study).

Physical chemistry of polymers (Bell Telephone Labs).

Project for plastics (Shell).

Questions

1. Classify the following changes according to the reaction type involved (e.g. addition, substitution, etc.) and name a suitable reagent in each case:

    (a) $CH_3 CHO \rightarrow CH_3 COOH$

    (b) $CH_3 CHO \rightarrow CH_3 CH(OH)CN$

    (c) $CH_3 CN \rightarrow CH_3 CH_2 NH_2$

    (d) $CH_3 CH_2 NH_2 \rightarrow CH_3 CH_2 OH$

    (e)

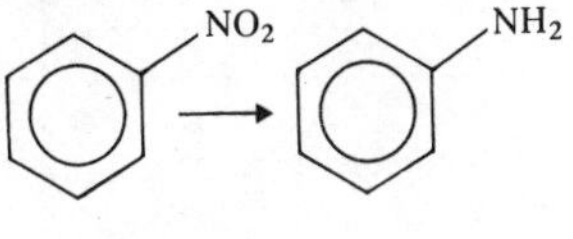

    (f)

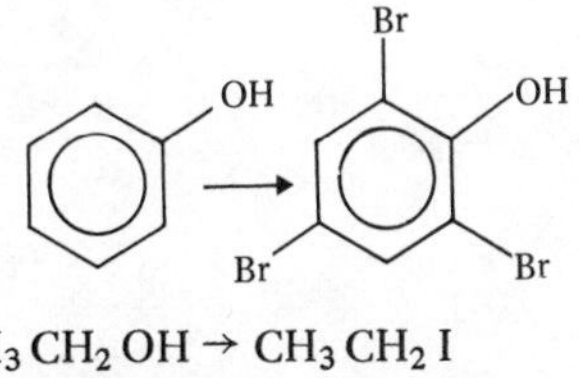

    (g) $CH_3 CH_2 OH \rightarrow CH_3 CH_2 I$

    (h) $(CH_3)_3 CI \rightarrow (CH_3)_2 C {=\!=} CH_2$

2. Describe how bromoethane is obtained: (a) in the laboratory; and (b) in industry.

How can this be converted into: (i) ethylbenzene; (ii) ethane; (iii) butane; (iv) ethanol; (v) ethyl ethanoate; (vi) propanoic acid.

3. Describe the behaviour of ethanoyl chloride and chloroethane with: (a) benzene; (b) ethanol; (c) ammonia; (d) water.

4. Distinguish between: (a) amines and amides; (b) addition polymers and condensation polymers; (c) the reactions of aldehydes and ketones with Grignard reagents.

5. How could you achieve the following five conversions in the laboratory?

(a) $CH_3CH_2COOH \rightarrow CH_3CH(NH_2)COOH$

(b) $CH_3CH_2OH \rightarrow CH_3CONH_2$

(c) CH_3COCl (and no other organic compound) $\rightarrow CH_3COOCH_2CH_3$

(d) $CH_3CHO \rightarrow CH_3CH(OH)CH_2NH_2$

(e)

6. How would you separate the following mixtures, isolating the compound indicated in a pure state?

(a) CH_3CHO from CH_3CH_2OH;

(b) C_6H_5COOH from aqueous ethoxyethane;

(c) $CH_3CH_2NH_2$ from CH_3CONH_2.

7. What are the characteristics of high polymers that make them industrially useful?

Nylon is a polyamide, Terylene is a polyester, Polythene is a poly-alkene. Discuss the meaning of these terms and suggest how the sub-stances might behave chemically.

Chapter 8

Mechanisms in organic chemistry

Background reading

The background material to this chapter is covered in *Fundamentals of Chemistry*, Chapters 2, 15 and 16.

Polarity of covalent bonds

Covalent bonds are formed by the sharing of pairs of electrons between two atoms. The atoms approach each other in bond formation until the unpaired electrons are attracted by both nuclei. For example, the electron in the 1s energy level of hydrogen is paired with the unpaired electron in the 2p energy level of fluorine (Fig. 8.1). The diagram shows the probability distribution

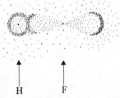

H F

Fig. 8.1 Combination of H (1s) and F (2p) electrons to form HF

around each atom; the bonding electrons must occupy the area of maximum probability for both atoms if a stable bond is to result (see *Fundamentals of Chemistry*, p. 23).

While this pair of electrons is attracted by both nuclei, the attractions are not necessarily of equal magnitude. The force acting on the electrons is determined by the distance of the electrons from the nucleus (the covalent radius; Table 8.1) and by the effective nuclear charge. The nucleus of fluorine has nine protons compared with the one proton of the hydrogen atom, but the nuclear attraction will not be nine times as great (even if the atoms were of the same size). This is because the nuclear effect is reduced by the inner electrons shielding the valence electrons from the full nuclear force.

Table 8.1 Some covalent radii

Atom	Covalent radius/pm
Br	114
C	77
Cl	99
F	72
H	37
I	133
Mg	136
N	74
O	74
S	104

The net effect of these two factors (distance and nuclear pull) is given in the term **electronegativity.** This is a measure of the net attractive force by an atom on a pair of electrons in a bond. Table 8.2 tabulates some common

Table 8.2 Some electronegativity values

Atom	Electronegativity
Br	2.8
C	2.5
Cl	3.0
F	4.0
H	2.1
I	2.5
Mg	1.2
N	3.0
O	3.5
S	2.5

values. A word of caution with respect to these values is necessary. The electronegativity of an atom (and its covalent radius too) depends on a number of factors, for example, the bond order, the bond length and the oxidation state.

This table shows that fluorine is more electronegative than hydrogen, i.e. it has a greater affinity for the electron pair. As a result, the electron density is greater on the fluorine than on the hydrogen. The bond between hydrogen and fluorine has become polar. The extent of this unequal sharing is determined by the nature of the atoms concerned (Fig. 8.2). In an extreme

	$\delta-$ $\delta+$	$-$ $+$
A : A	A : B	A: B
Non-polar	Polar	Ionic
(equal sharing)	(unequal sharing)	(complete transfer)

Fig. 8.2 Electron pair distribution between two atoms

situation, the electron pair is effectively associated with one atom only. The unequal sharing of the electrons results in a charge distribution as shown.

It is convenient in organic chemistry to compare the polarity of a bond to that of the C—H bond in the related alkane. The polarisation of the bond relative to the C—H bond is described as the **inductive effect**. A group (e.g. a halogen) attached to a carbon framework may be 'electron attracting', being more electronegative than carbon:

$\delta+$ $\delta-$
C–Cl

Groups of this type, which involve a displacement of the electrons away from the carbon, are described as electron-withdrawing groups. The electron density around this group is greater than that around a hydrogen atom in an otherwise analogous structure, and it is said to have a negative inductive effect:

$\sim\sim$ C $\rightarrow$ X	$-I$ effect
$\sim\sim$ C —— H	reference
$\sim\sim$ C $\leftarrow$ X	$+I$ effect

If a group is less electron attracting than hydrogen, the electron density will increase around the carbon atom relative to the analogous C—H structure. Such a group is described as possessing a positive inductive effect. Table 8.3

Table 8.3 Inductive effects

+ I effect	—I effect
O^-	NR_3^+
$CO \cdot O^-$	NO_2
$C(CH_3)_3$	CHO
$CH(CH_3)_2$	F, Cl, Br, I
CH_2CH_3	OH, OR
	NH_2

lists groups displaying these effects. It will be noticed that the $-I$ effect is displayed by positively charged groups, groups possessing $=\!\!=\!\!O$ bonds, and atoms or groups which are more electronegative than hydrogen. Conversely, negatively charged groups and alkyl groups (less electronegative than hydrogen) display $+I$ effects.

Strengths of organic acids

This inductive effect can be demonstrated in a number of examples. Table 8.4 lists the pK_a values for a number of acids. If one hydrogen of ethanoic acid is replaced by a chlorine atom, the pK_a drops (the acid strength has increased). This indicates that the hydrogen atom of the carboxylic group is more easily released. Since the chlorine is an electron-withdrawing group, electrons are removed from the carboxylic grouping, reducing the electron density in the O—H bond, thereby facilitating the removal of H^+ by the solvent molecules:

$$Cl \twoheadleftarrow CH_2 \twoheadleftarrow \overset{\overset{\displaystyle O}{\|}}{C} \twoheadleftarrow O \twoheadleftarrow H$$

The effect is accentuated by introducing two or three chlorine atoms. Fluorine has a stronger effect than chlorine; bromine and iodine display weaker effects. This is in accord with the electronegativity values (Table 8.2). Several other groups are listed which display the effects predicted from Table 8.3. It is noted that the alkyl groups weaken the acid strength.

Table 8.4 pK_a values for some organic acids $R \cdot COOH + H_2O \rightleftharpoons RCOO^- + H_3O^+$

Formula	pK_a (298 K)
$H-CH_2-COOH$	4.76
$Cl-CH_2-COOH$	2.86
$Cl_2CH-COOH$	1.29
$Cl_3C-COOH$	0.65
$F-CH_2-COOH$	2.66
$Br-CH_2-COOH$	2.90
$I-CH_2-COOH$	3.17
$HO-CH_2-COOH$	3.83
$NC-CH_2-COOH$	2.47
O_2N-CH_2-COOH	1.68
$HOOC-CH_2-COOH$	2.83
CH_3CO-CH_2-COOH	3.58
CH_3-CH_2-COOH	4.87
$(CH_3)_3C-COOH$	5.05

The inductive effect reduces in intensity along a carbon chain. The further a group is from the centre of reactivity, the less its effect on the properties of the compound. This is illustrated in Table 8.5. It will be seen

Table 8.5 Comparison of inductive effects as chain length increases

Acid	pK_a	Acid	pK_a	Acid	pK_a
$ClCH_2COOH$	2.86	$ClCH_2CH_2COOH$	3.99	$ClCH_2CH_2CH_2COOH$	4.52
$BrCH_2COOH$	2.90	$BrCH_2CH_2COOH$	4.00	$BrCH_2CH_2CH_2COOH$	4.59
ICH_2COOH	3.15	ICH_2CH_2COOH	4.05	$ICH_2CH_2CH_2COOH$	4.64
$NCCH_2COOH$	2.47	$NCCH_2CH_2COOH$	3.99	$NCCH_2CH_2CH_2COOH$	4.43

that the inductive effect drops off by a factor of approximately 60 per cent for each extra carbon atom in the chain.

Strength of bases

Base strength is affected in a similar manner. Table 8.6 gives the strength of nitrogen bases in terms of pK_a and pK_b values; the stronger bases have low pK_b and high pK_a values. The base strength is measured by the ability of the compound to abstract a proton from water molecules:

$$R\!-\!\underset{\underset{H}{|}}{\overset{\overset{H}{|}}{N}}: + H_2O \rightleftharpoons R\!-\!\overset{+}{\underset{\underset{H}{|}}{\overset{\overset{H}{|}}{N}}}\!-\!H \ + OH^-$$

$$K_b = \frac{[R\overset{+}{N}H_3]\,[OH^-]}{[RNH_2]}$$

If a substituent group is electron withdrawing, the base strength is reduced because of its lesser ability to form a donor bond with the proton:

$$-N: + H^+ \rightarrow N:H$$

The alkyl group exerts a +I effect and so, compared to ammonia, the amino-alkanes have increased donor ability and increased base strength. The carbonyl group (CH_3CO—) in ethanamide is electron-withdrawing (−I effect) and so the amine group is less basic in the amide than in an amine (see p. 196).

Table 8.6 Strengths of some organic bases (pK_b) and their conjugate acids (pK_a)

Base	pK_a	pK_b
$H\!-\!NH_2$	9.25	4.75
CH_3NH_2	10.64	3.36
$CH_3CH_2NH_2$	10.93	3.07
CH_3CONH_2	−1.1	15.1
$(CH_3)_2NH$	10.72	3.28

Reactivity of acyl halides

Halogenoalkanes react with alcohols (in the presence of a strong base) to give ethers:

$$R \cdot OH + Na \rightarrow RO^-Na^+$$

$$R\!-\!\overset{-}{O}: + \underset{\delta+}{CH_2}\!\!\underset{\delta-}{\rightarrow\!\!Cl} \ \rightarrow \ R\!-\!O\!-\!CH_2\!\!-\!\!CH_3 \ + \ Cl^-$$

(The curved arrows indicate the movement of electrons between atoms or molecules.) Acyl halides react much more readily with alcohols (p. 196). This can be related to the inductive effect. From the above reaction scheme it is seen that the alcoholic reagent attacks the positively polarised carbon atom of the halogenoalkane. In ethanoyl chloride, the carbon atom is polarised due to both the chlorine atom and the carbonyl group, since both exert $-I$ effects. In this case it is not necessary to accentuate the polarity of the alcohol by adding sodium metal.

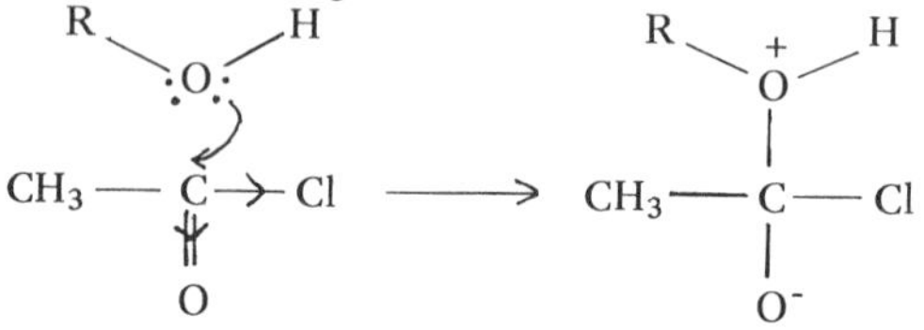

$$CH_3 - \overset{\overset{\displaystyle R\diagdown \;\; \diagup H}{\underset{..}{O:}}}{\underset{\underset{\displaystyle O}{\|}}{C}} \rightarrow Cl \longrightarrow CH_3 - \overset{\overset{\displaystyle R\diagdown \overset{+}{} \diagup H}{O}}{\underset{\underset{\displaystyle O^-}{|}}{C}} - Cl$$

An unstable intermediate is formed by the movement of one pair of electrons from the alcoholic oxygen to the carbon, and one pair from the carbon to the carbonyl oxygen. This intermediate then expels the Cl^- ion and undergoes deprotonation to yield an ester:

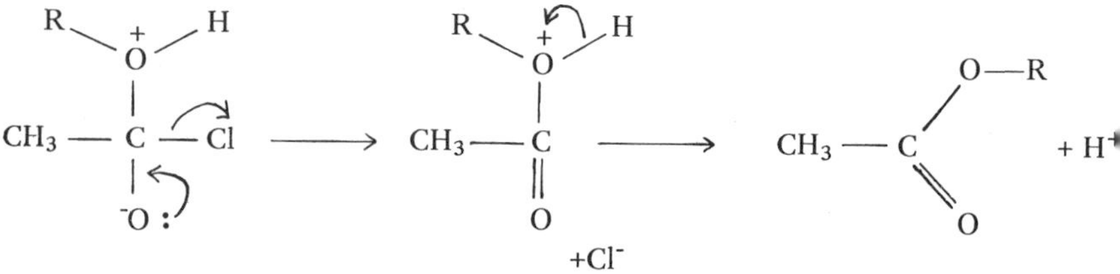

$$CH_3 - \overset{\overset{R\diagdown \overset{+}{O} \diagup H}{|}}{\underset{\underset{^-O:}{|}}{C}} - Cl \longrightarrow CH_3 - \overset{\overset{R\diagdown \overset{+}{O} \diagup H}{|}}{\underset{\underset{O}{\|}}{C}} \longrightarrow CH_3 - C\overset{\diagup O - R}{\diagdown_{O}} + H^+$$

$$+Cl^-$$

The effect of multiple bonds

The inductive effect is applicable to multiply bonded systems (e.g. carbonyls) as well as to singly bonded compounds (e.g. alcohols). In the former, the π-bonds are more easily distorted than the σ-bond and so the effects are accentuated (Fig. 8.3). The inductive effect in sigma bonds is usually repre-

Fig. 8.3 Polarisation of a π-bond

sented as an arrow on the bonded line ($C\rightarrow O$), but the effect in the π-bond is shown by the curved arrows.

Delocalisation of electrons

When carboxylic acids react with water molecules, carboxylate ions are formed:

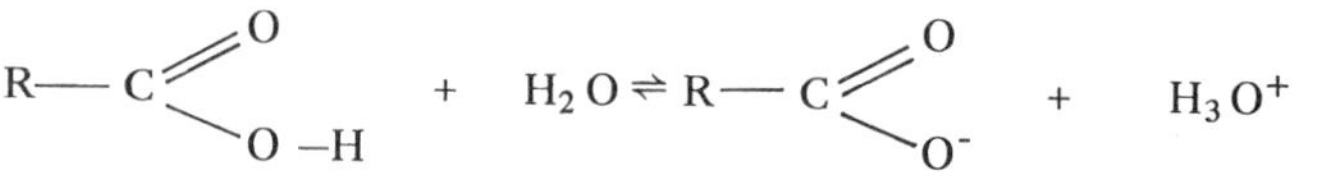

$$R - C\overset{\diagup O}{\diagdown_{O - H}} + H_2O \rightleftharpoons R - C\overset{\diagup O}{\diagdown_{O^-}} + H_3O^+$$

The nominal structure shown involves two types of carbon–oxygen bond, $C{=}O$ and $C{-}O^-$. By comparison with the parent molecule, it would be expected that these bonds would be 124 and 143 pm long respectively. In fact, they are of equal length, viz. 127 pm. Furthermore, the negative charge is shared equally between the two oxygen atoms and the molecule is symmetrical:

$$R{-}C\underset{O_{\frac{1}{2}}-}{\overset{O_{\frac{1}{2}}-}{\lessgtr}}$$

While this effect can be related to the inductive effects in the two bonds, it is greater than would be anticipated by them.

Electronically, this phenomenon in which the electrons are delocalised over several atoms can be interpreted as follows. The π-bond is formed by the overlap of the p_z orbitals on the carbon and the oxygen of the carbonyl group (Fig. 8.4a). This is polarised due to electronegativity differences. The positively polarised carbon atom now has a strong effect on the negatively charged oxygen ($C{-}O^-$ bond) and attracts the p_z electrons of this atom into its partially vacated orbital. The internuclear distance is decreased to give better overlap of the orbitals, the energy release from this compensating for the increase in internuclear repulsion forces.

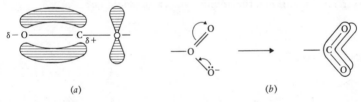

(a) (b)

Fig. 8.4 Structure of the carboxylate ion

In this manner, the electron density (and so the negative charge) is distributed over the whole group (Fig. 8.4b). In general, if the electron density can be more widely distributed by this means, greater stability results.

This phenomenon occurs widely in organic chemistry. A similar effect to that described for carboxylates occurs in unsaturated amines:

$$\overset{..}{N}{-}C{=}C$$

and unsaturated alcohols (enols):

$$\overset{..}{O}{-}C{=}C$$
H

In the former case the basicity decreases (since the nitrogen is less able to donate electrons to a proton). Unsaturated alcohols are more acidic than their saturated analogues, because the reduced electron density on the oxygen atom facilitates the removal of a proton.

Compounds involving multiply bonded groups (e.g. carbonyls) display the reverse effect, because of the electronegativity of the atom concerned:

but the result is the same in that the π-electrons are delocalised:

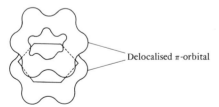

The effect is termed **mesomerism.**

In these examples, it has been shown that compounds which contain multiple bonds in alternation with single bonds (i.e. the C=C—C=C arrangement) undergo an extensive delocalisation of the π-electrons (e.g. C==C==C==C). This effect is known as **conjugation.** The presence of conjugation enables a molecule to extend the influence of a substituent through the chain of carbon atoms more efficiently than occurs with the inductive effect (p. 221). The effect is important, not only in the conjugated dienes and unsaturated carbonyl compounds shown above, but also in aromatic systems. In benzene, the nominal structure would be

H
H   H
H   H
H

Conjugation of the π-bonds gives an extensively delocalised system (Fig. 8.5).

Delocalised π-orbital

Fig. 8.5   Delocalisation of the π-bonds in benzene

Reagents

Organic reagents can be classified into two main classes: nucleophilic and electrophilic reagents. A smaller, though not unimportant group involves free radicals.

The **nucleophilic** reagents are those possessing full or partial negative charge (e.g. Br^-, OH^-, CN^-, H_2O, ROH, NH_3, RNH_2) and are, therefore, attracted to a positive centre (i.e. nucleophilic or 'nucleus loving', based on the Greek). They are Lewis bases (*Fundamentals of Chemistry*, p. 29). They attack, for example, positively charged carbon atoms of halogenoalkanes:

$$\overset{-}{N}C : \ + \ \overset{\delta+}{C}\overset{\delta-}{\underset{}{—X}} \ \rightarrow \ NC—C \ + \ X^-$$

Electrophilic reagents (e.g. H^+, Br^+, RN_2^+, NO_2^+) are positively charged Lewis acids and attack negative centres (so 'electron loving'). A typical example is the nitration of benzene:

$$+ \ NO_2^+ \ \longrightarrow \ + \ H^+$$

Reaction types

Substitution reactions

1. Nucleophilic substitutions

The halogen atom of a halogenoalkane can be displaced by a nucleophile. For example, in alkaline hydrolysis:

$$H\overset{-}{O} : \ + \ \overset{\delta+}{C}\overset{\delta-}{\rightarrow Br} \ \rightarrow \ HO—C \ + \ Br^-$$

or, in the reaction with ammonia:

$$H_3N : \ + \ \overset{\delta+}{C}\overset{\delta-}{\rightarrow Br} \ \rightarrow \ H_3\overset{+}{N}—C \ + \ Br^-$$

The order of reactivity of the halogenoalkanes is determined by the bond strength. The iodoalkanes are the most reactive of the halogenoalkanes since the bond energy is least (Table 8.7). As the bond length increases, the bond energy decreases and so the reactivity increases.

Table 8.7 Some typical bond energies

Bond type	Bond energy/kJ mol^{-1}	Bond length/pm
C—F	485	138
C—Cl	338	177
C—Br	284	194
C—I	213	214

The bimolecular step involves the formation of a transition state with a five-bonded carbon. Clearly this is an unstable arrangement, but it is achieved by the partial formation of the new bond together with the weakening of the carbon–halogen bond (Fig. 8.6). The transition state is very unstable and

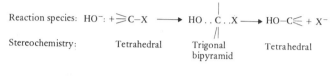

Reaction species: HO^-: $+ \,\rangle C{-}X \longrightarrow HO..C..X \longrightarrow HO{-}C\langle + X^-$

Stereochemistry: Tetrahedral   Trigonal   Tetrahedral
                                        bipyramid

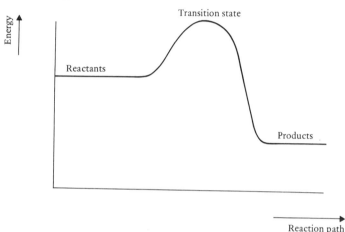

Fig. 8.6   Reaction profile for an S_N2 reaction

is rapidly broken down. The formation of this species is slow, being hindered sterically and electronically. The steric effect is due to the difficulty that the reagent has in reaching a quarternary substituted carbon atom; four groups are distributed around the atom in a tetrahedral arrangement. The electronic configuration of the transition state is also unfavourable. It implies the existence of five bonds on the carbon atom; this atom does not have sufficient atomic orbitals to form five bonds. So, the overall rate is determined by the first, bimolecular step.

HO^- + $\,\rangle C{-}Br \rightarrow [HO \ldots C \ldots Br]^-$ slow

$[HO \ldots C \ldots Br]^- \rightarrow HO{-}C\langle + Br^-$ fast

The reaction mechanism is described by the nature of the slower or rate-determining step as a bimolecular nucleophilic substitution reaction (S_N2). The rate expression for this mechanism indicates that it is first order with respect to each reagent:

rate = $k \cdot [HO^-] \cdot [RX]$

The formation of this transition state is most likely for the primary alkyl species ($RCH_2 X$) in which the carbon is more polar than in the more highly substituted species:

$$
\begin{array}{ccc}
\text{H} & \text{H} & \text{R} \\
| & | & \downarrow \\
\text{R}\rightarrow\text{C}\rightarrow\text{X} \quad & \text{R}\rightarrow\text{C}\rightarrow\text{X} \quad & \text{R}\rightarrow\text{C}\rightarrow\text{X} \\
| & \uparrow & \uparrow \\
\text{H} & \text{R} & \text{R}
\end{array}
$$

The increased number of alkyl groups reduces the polarity of the carbon atom because of their electron-donating effect. Also, as the more bulky alkyl groups replace the hydrogen atoms, attack by the reagent is hindered. Since the transition state involves a delocalised charge, a solvent of low polarity (i.e. of small dielectric constant) is favoured.

The alternative mechanism involves a monomolecular nucleophilic substitution ($S_N 1$) for the rate-determining step:

$$
\begin{array}{ccc}
\text{R} \\
\quad\searrow \\
\text{R}-\text{C}-\text{X} \quad\rightarrow\quad \overset{\text{R} \quad \text{R}}{\underset{\text{R}}{\searrow\text{C}^+\swarrow}} \quad + \quad \text{X}^- & \qquad \text{slow} \\
\quad\nearrow \\
\text{R}
\end{array}
$$

$$
\text{HO}^- \quad + \quad \overset{\text{R} \quad \text{R}}{\underset{\text{R}}{\searrow\text{C}^+\swarrow}} \quad \rightarrow \quad \overset{\text{R}}{\underset{\text{R}}{\text{HO}-\text{C}-\text{R}}} \qquad \text{fast}
$$

In this mechanism, the reaction is initiated by the breaking of the carbon–halogen bond to give a carbonium ion intermediate (Fig. 8.7). Carbonium ions are species with positively charged carbon atoms.

The rate expression for these reactions is

rate $= k$ [RX]

The rate is independent of the concentration of alkali. (The concentration of alkali may affect the yield but not the rate.)

Since the intermediate is ionic, a polar solvent is desirable to facilitate the reaction. Steric problems are of less significance in reactions proceeding by the $S_N 1$ route than in those following the $S_N 2$ mechanism. This is because the carbonium ion has a large region open to attack by the nucleophile. The formation of the carbonium ion is facilitated if the ion can be stabilised. Tertiary substituted carbons form more stable carbonium ions than primary groups because the positive charge is reduced due to the +I effects of the alkyl groups. A lower charge reduces the polarising power of the ion, so making it easier for the halogen to break free from the carbon atom.

The alternative reaction of elimination which is also possible has not

Reaction species: $R_3C-X + HO^- \longrightarrow R_3C^+ + OH^- \longrightarrow R_3C-OH + X^-$
$+ X$

Stereochemistry: Tetrahedral   Trigonal   Tetrahedral

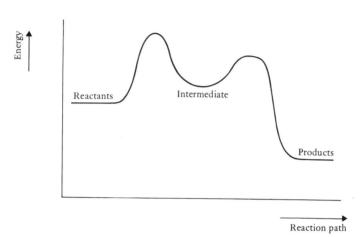

Fig. 8.7   Reaction profile for an S_N1 reaction

been considered (see p. 189). This reaction competes with substitution particularly in the case of the S_N1 reactions.

2. Electrophilic substitution

For an electrophile to substitute at a carbon atom, the carbon must be a centre of negative charge. This occurs most readily with aromatic compounds. Benzene is susceptible to electrophilic attack because of the π-electron density which attracts such reagents. For example, the nitration of benzene proceeds as follows:

The positive charge on the nitronium ion (NO_2^+) attracts the π-electrons of the benzene. This causes the bond to break at one position to form a positively charged intermediate (Fig. 8.8). This intermediate has the

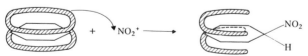

Fig. 8.8   Nitration of benzene

π-electrons distributed over five carbon atoms and the positive charge is also spread over these atoms.

The nitronium ion, NO_2^+, is present in small concentrations in nitric acid; its formation is accentuated by the presence of concentrated sulphuric acid. This acid shifts the equilibrium in favour of the nitronium ion by the removal of water:

$$H_2SO_4 + HNO_3 \rightleftharpoons HSO_4^- + H_2\overset{+}{O}-NO_2$$
$$\Updownarrow$$
$$H_2O + NO_2^+$$

A number of neutral molecules, or those of low polarity, can also be used by inducing polarisation. Bromine is polarised by the presence of a Lewis acid such as iron(III) bromide. Iron may be used, but this consumes some of the reagent for the formation of the bromide:

$$2Fe + 3Br_2 \rightarrow 2FeBr_3$$

$$Br_2 + FeBr_3 \rightleftharpoons \overset{\delta+}{Br} \overset{\delta-}{—BrFeBr_3}$$

The polarised species then reacts with benzene:

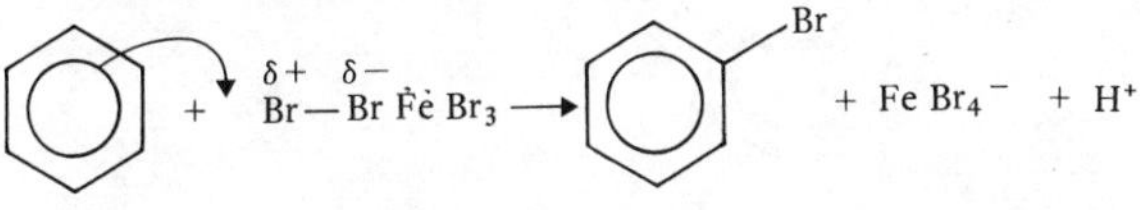

Aluminium chloride is used as a catalyst in the Friedel–Crafts' reaction:

$$2CH_3Cl + Al_2Cl_6 \rightleftharpoons 2CH_3\overset{\delta+}{—}\overset{\delta-}{ClAlCl_3}$$

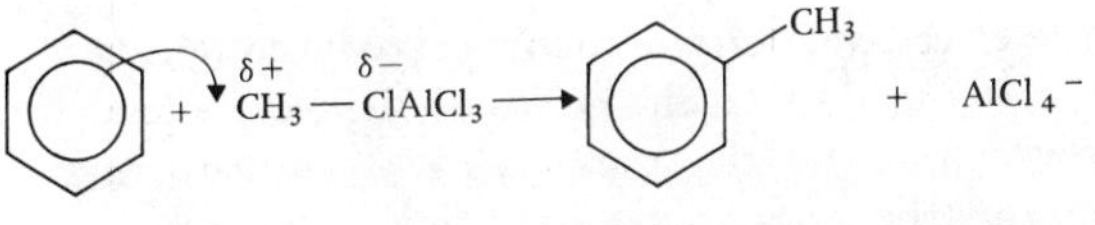

As shown in the previous chapter, substituted benzene compounds may be activated or deactivated towards further electrophilic substitution according to the nature of the first substituent. The activating groups (—OH, —CH$_3$, —NH$_2$ etc.) push electrons into the ring through the π-bond system (compare p. 198). The electron-pushing effect causes a build-up of charge in the 2, 4 and 6 positions:

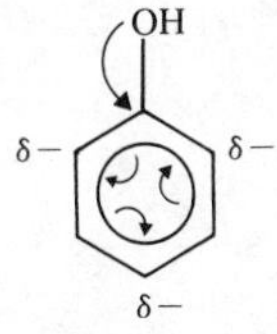

These positions are then more susceptible to electrophilic attack than are the carbon atoms of benzene. So, whereas nitration of benzene is slow, nitration of phenol is faster. Also a less strong nitrating agent is sufficient; sodium nitrate–dilute sulphuric acid mixture attacks phenol satisfactorily. Bromination of phenol occurs so readily that it may be achieved with an aqueous solution of bromine and results in the trisubstituted product:

Similar results occur with aniline (aminobenzene). The —OH and —NH$_2$ groups exert small —I inductive effects but the delocalisation effect is much greater and results in a net donation of electrons to the ring.

Methylbenzene (toluene) has only an inductive effect, but this also results in electron donation to the benzene ring. So, in the Friedel–Craft's reaction, the alkylbenzene produced activates the ring towards further attack. Ultimately a hexamethylbenzene is produced.

Groups such as the aldehydes (—CH=O), nitro (—N$\overset{O}{\underset{O}{<}}$), and carboxylate (—C$\overset{O}{\underset{OH}{<}}$) have an electron-withdrawing effect through delocalisation. In these compounds the electron density in the ring is reduced (giving less susceptibility to electrophilic attack) but the positions 3 and 5 are less affected than the other carbon atoms. So, the reactivity of the aromatic ring is reduced compared to benzene (the groups are **deactivating**) and the 3-substituted product is more probable. The decrease in reactivity is illustrated by the reaction conditions required. The maximum amount of mononitration is achieved in about 45 minutes at below 50 °C; dinitration

is achieved by heating the mixture to above this temperature; trinitration requires a large excess of the nitrating mixture and a temperature of over 100 °C maintained for several days.

The deactivation of the ring in chlorobenzene is due to the large inductive effect reducing the electron density in the ring. The mesomeric effect operates in the opposite direction, but is weaker. It is, however, enhanced on the approach of an electrophilic reagent (e.g. NO_2^+ in nitration) and so it gives substitution in the 2 and 4 positions as expected of an

electron donor (see above). This enhanced mesomeric effect is sometimes known as the **electromeric effect.** Table 8.8 gives the relative rates of

Table 8.8   Relative rates of nitration

Compound	Relative rate
Aminobenzene	$\sim 10^6$
Methylbenzene	25
Benzene	1
Chlorobenzene	0.033
Nitrobenzene	$\sim 10^{-6}$

nitration of activated and deactivated compounds.

Addition reactions

1. Electrophilic reactions

Electrophilic addition reactions occur in a manner similar to the electrophilic substitution reactions of benzene. When addition to an alkene occurs, the π-electrons undergo electrophilic attack.

$$\begin{array}{c} CH_2 \\ \| \\ CH_2 \end{array} + \overset{\delta+}{H}-\overset{\delta-}{Br} \longrightarrow \begin{array}{c} {}^+CH_2 \\ | \\ CH_3 \end{array} + {}^{\cdot\cdot}Br^- \longrightarrow \begin{array}{c} CH_2-Br \\ | \\ CH_3 \end{array}$$

Homonuclear molecules such as bromine are able to attack without catalytic activation since the π-electrons induce some polarisation of the covalent bond.

$$\begin{array}{c} CH_2 \\ \| \\ CH_2 \end{array} + Br-Br \longrightarrow \begin{array}{c} CH_2 \\ | \\ CH_2 \end{array}\!\!Br^+ + Br^-$$

$${}^-Br\!: + \begin{array}{c} CH_2 \\ | \\ CH_2 \end{array}\!\!Br^+ \longrightarrow \begin{array}{c} Br-CH_2 \\ | \\ CH_2-Br \end{array}$$

If alternative negative polar species are present (e.g. Cl^-, H_2O), then the second step may result in mixed products:

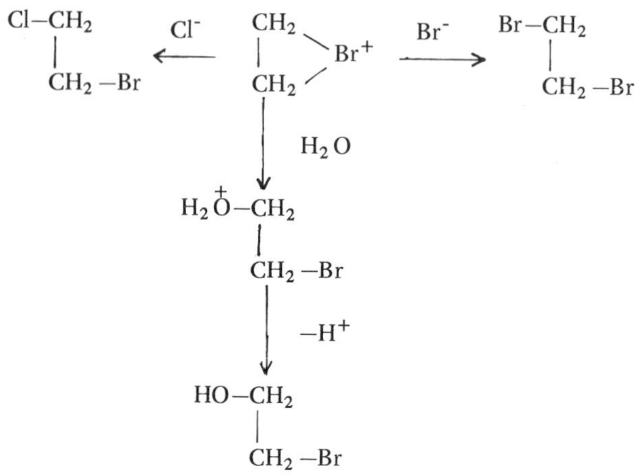

$$Cl-CH_2 \quad \xleftarrow{\;Cl^-\;} \quad \begin{array}{c} CH_2 \\ | \\ CH_2 \end{array}\!\!\diagdown\!Br^+ \quad \xrightarrow{\;Br^-\;} \quad Br-CH_2$$

$$\begin{array}{c} | \\ CH_2-Br \end{array} \qquad\qquad\qquad\qquad\qquad \begin{array}{c} | \\ CH_2-Br \end{array}$$

$$\Big\downarrow H_2O$$

$$H_2\overset{+}{O}-CH_2$$
$$|$$
$$CH_2-Br$$

$$\Big\downarrow -H^+$$

$$HO-CH_2$$
$$|$$
$$CH_2-Br$$

2. Nucleophilic reactions

Addition of nucleophilic reagents requires a polarised multiply bonded system such as the carbonyl group.

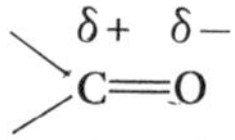

$$\diagdown\!\!\overset{\delta+\;\;\delta-}{C=O}$$

A typical polar reagent is hydrogen cyanide which reacts with aldehydes in the presence of a basic catalyst.

$$HCN + OH^- \rightarrow H_2O + CN^-$$

The cyanide ion attacks the polar carbon atom

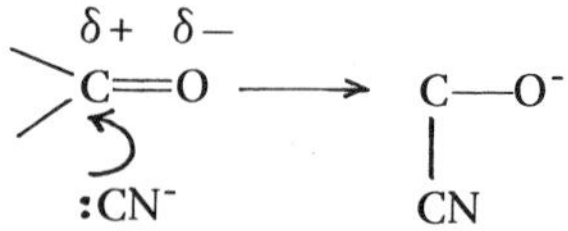

$$\diagdown\!\!\overset{\delta+\;\;\delta-}{C=O} \longrightarrow \begin{array}{c} C-O^- \\ | \\ CN \end{array}$$
$$:CN^-$$

and the basic product removes a proton from more HCN:

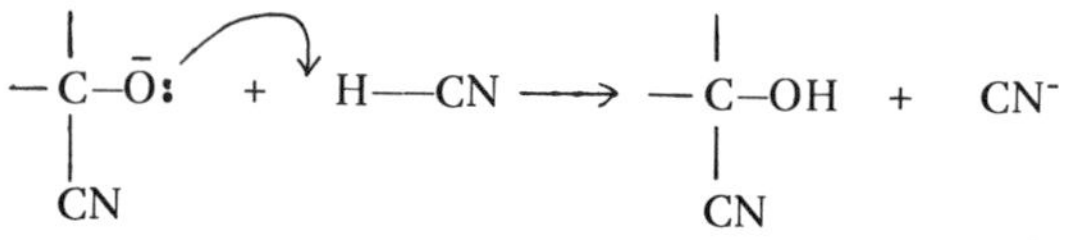

$$\begin{array}{c} | \\ -C-\bar{O}: \\ | \\ CN \end{array} + \;H-CN \longrightarrow \begin{array}{c} | \\ -C-OH \\ | \\ CN \end{array} + \; CN^-$$

The positive charge on the carbon is reduced in the series

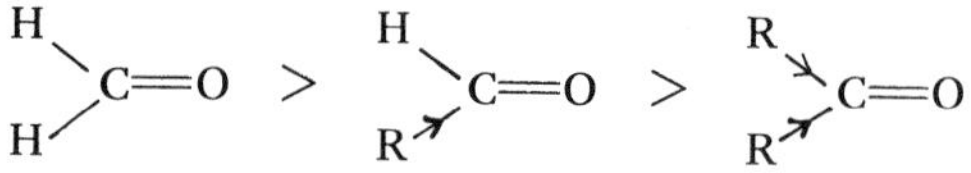

$$\overset{H}{\underset{H}{\diagup}}\!\!\diagdown\!C=O \;>\; \overset{H}{\underset{R}{\diagup}}\!\!\diagdown\!C=O \;>\; \overset{R}{\underset{R}{\diagup}}\!\!\diagdown\!C=O$$

due to the inductive effect of the alkyl groups, so that ketones are usually less reactive than aldehydes.

Nucleophilic attack also occurs between carbonyls and hydrogen sulphite ions (HSO_3^-):

$$\underset{\text{:}SO_3H^-}{\overset{\delta+\;\;\delta-}{>C=O}} \quad\rightleftharpoons\quad \underset{SO_3H}{\overset{|}{-\overset{|}{C}-\overset{\bar{}}{O}\text{:}}} \quad\rightleftharpoons\quad \underset{SO_3^-}{\overset{|}{-\overset{|}{C}-OH}}$$

and hydride ions (as complexes such as AlH_4^-):

$$\underset{\text{:}H^-}{\overset{\delta+\;\;\delta-}{>C=O}} \quad\longrightarrow\quad \underset{H}{\overset{|}{-\overset{|}{C}-O^-}} \quad\overset{H_2O}{\longrightarrow}\quad \underset{H}{\overset{|}{-\overset{|}{C}-OH}} + OH^-$$

Carbonyls can be activated towards nucleophiles if an electron-withdrawing group is used:

$$\underset{R}{\overset{Cl}{>}}C=O \quad > \quad \underset{R}{\overset{H}{>}}C=O$$

An acid chloride is susceptible to attack by water or alcohols:

$$\underset{\text{:}OH_2}{\overset{Cl}{R-\overset{|}{C}=O}} \quad\rightleftharpoons\quad \underset{\overset{OH_2}{+}}{\overset{Cl}{R-\overset{|}{\underset{|}{C}}-O^-}} \quad\rightarrow\quad \underset{OH}{R-\overset{|}{C}=O} + Cl^- \\ + H^+$$

and by ammonia

$$\underset{\text{:}NH_3}{\overset{Cl}{R-\overset{|}{C}=O}} \quad\rightleftharpoons\quad \underset{\overset{NH_3}{+}}{\overset{Cl}{R-\overset{|}{\underset{|}{C}}-O^-}} \quad\rightarrow\quad \underset{NH_2}{R-\overset{|}{C}=O} + Cl^- \\ + H^+$$

Electrophilic attack on the oxygen is only significant in the presence of a strong acid and it generates a carbonium ion:

$$\underset{}{\overset{OH}{R-\overset{|}{C}=O\text{:}}} + H^+ \quad\rightleftharpoons\quad \underset{+}{\overset{OH}{R-\overset{|}{C}-OH}}$$

This ion can then react with a nucleophile (e.g. an alcohol, giving an ester).

The so-called condensation reactions with ammonia, or its substituted derivatives, involve an addition reaction as a key step:

$$R-\underset{\underset{:NH_2X}{\displaystyle |}}{\overset{\overset{\displaystyle H}{|}}{C}}\overset{\delta+\ \delta-}{=O} \ \rightleftharpoons \ R-\underset{\underset{\underset{+}{NH_2X}}{\displaystyle |}}{\overset{\overset{\displaystyle H}{|}}{C}}-O^- \ \rightleftharpoons \ R-\underset{\underset{:NHX}{\displaystyle |}}{\overset{\overset{\displaystyle H}{|}}{C}}-OH$$

$$H^+ \ + \ R-\underset{\underset{NX}{\|}}{\overset{\overset{\displaystyle H}{|}}{C}} \ \rightleftharpoons \ R-\underset{\underset{\underset{+}{NHX}}{\|}}{\overset{\overset{\displaystyle H}{|}}{C}} \ + OH^-$$

Elimination reactions

1. Dehydrobromination

It was mentioned above (p. 189) that an elimination reaction could occur as an alternative mechanism to substitution. Elimination results in the formation of an alkene:

$$CH_3CH_2Br \ + \ OH^- \overset{substitution}{\underset{elimination}{\Big\langle}} \begin{array}{l} CH_3CH_2OH \\[1em] CH_2{=}CH_2 \ + Br^- \\ \qquad\qquad + \ H_2O \end{array}$$

In the case of the primary alkyl derivatives, the yield is very low (less than 1%) even under the most favourable conditions (see below). Conversely, tertiary alkyl halides yield virtually 100 per cent alkene.

The basic reaction involves attack by the nucleophile on a 'β-hydrogen', that is a hydrogen atom on the second (or β) carbon atom from the substituent:

$$HO\overset{-}{:} \ + \ H{\rightarrow}CH{\rightarrow}CH_2{\rightarrow}X \ \overset{slow}{\longrightarrow} \ [HO \ldots H \ldots \overset{\overset{\displaystyle R}{|}}{CH} \cdots CH_2 \ldots X]$$

transition compound

$$\downarrow fast$$

$$H_2O + R \cdot CH{=}CH_2 + X^-$$

The mechanism shown is a bimolecular elimination (E2) reaction. Alternatively, an E1 mechanism (unimolecular elimination) might be operative, involving a carbonium ion:

$$(CH_3)_3C-Br \ \rightarrow \ (CH_3)_3C^+ \quad slow$$

intermediate

$$(CH_3)_2\overset{+}{C}\text{---}CH_2\text{---}H \quad + \quad :\overline{O}H \longrightarrow \quad (CH_3)_2C\text{=}CH_2 \ + \ H_2O \quad \text{fast}$$

As in the corresponding substitution reactions, the E1 reaction rate is independent of alkali concentration; the E2 reaction is dependent on it. The factors affecting the substitution reaction mechanism (solvent, steric, inductive effect of alkyls) affect the choice of elimination mechanism in the same way. Several factors are important in favouring elimination or substitution.

Elimination is favoured against substitution as the degree of branching of the alkyl group is increased. Tertiary halides are less susceptible to substitution attack by nucleophiles because of the reduced charge on the carbon atom. Steric factors are also important — large groups around the polar carbon hinder attack by the reagent. Abstraction of hydrogen is aided if there is a positive charge on the carbon (removing electron density from the hydrogen) and if a strong base (rather than a weaker one) is used. Base strength decreases in the order

$$NH_2^- > OR^- > OH^- > O \cdot COCH_3^- > CN^-$$

A concentrated solution of the base will also favour elimination. Since the elimination reaction generally has a higher activation energy than the competing substitution reaction, the proportion of the former type can be increased by raising the temperature. Table 8.9 indicates the conditions that favour alkene formation against nucleophilic substitution. However, it should be noted that

Table 8.9 Conditions for elimination and substitution reactions

Conditions	Reaction favoured	
	Elimination	Substitution
Alkyl group	Tertiary	Primary
Steric effect	Branched chain	Small, unbranched chain
Solvent	Polar	Non-polar
Base	Strong	Weak
Temperature	Hot solution	Warm solution
Most likely mechanism	E1	S_N2

primary halides with strong bases (e.g. alcoholic KOH) at high temperature still give very little alkene; tertiary halides with weak bases (e.g. CN^-) tend to give the alkene almost exclusively. The primary halides undergo bimolecular substitution, S_N2 (p. 226), whereas a monomolecular elimination (E1) occurs with tertiary halides (p. 234).

2. Dehydration

The most common example of a dehydration reaction is that involving the action of concentrated sulphuric acid on an alcohol. Two types of reaction arise — one when the acid is in excess and the other with excess alcohol (both requiring high temperature). In each case the first step is protonation of the alcohol:

$$CH_3 CH_2 OH + H^+ \rightleftharpoons CH_3 CH_2 \overset{+}{O}H_2$$

(a) Excess acid – hydrogen sulphate anions abstract a proton from the β-carbon to form an alkene:

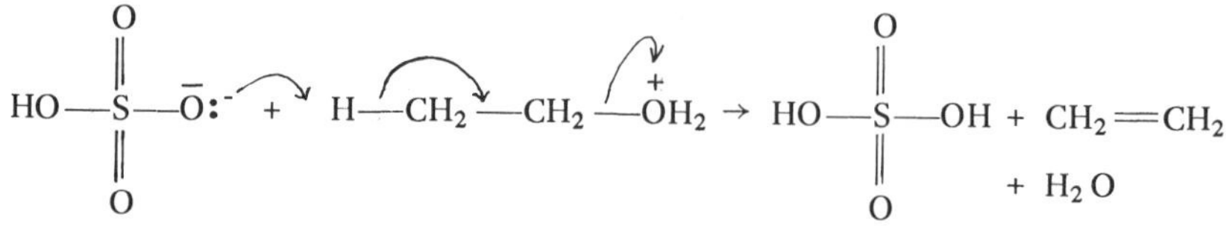

(b) Excess alcohol – the alcohol acts as a nucleophilic reagent, and a substitution reaction occurs giving an ether:

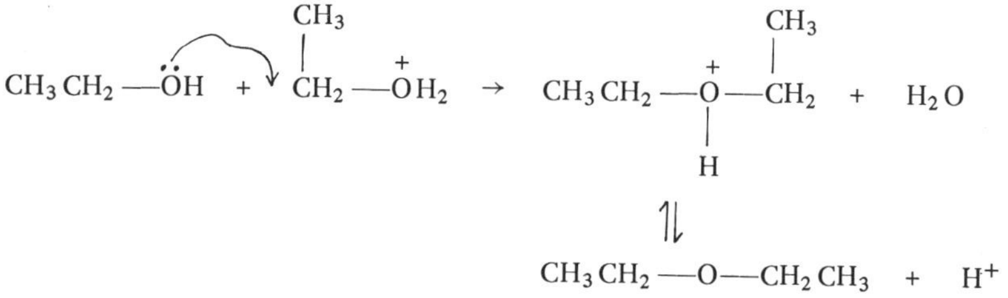

3. Deamination

Amine groups are displaced from quaternary ammonium salts by the action of hot base to produce an alkene. As with the other elimination reactions, a β-carbon is essential.

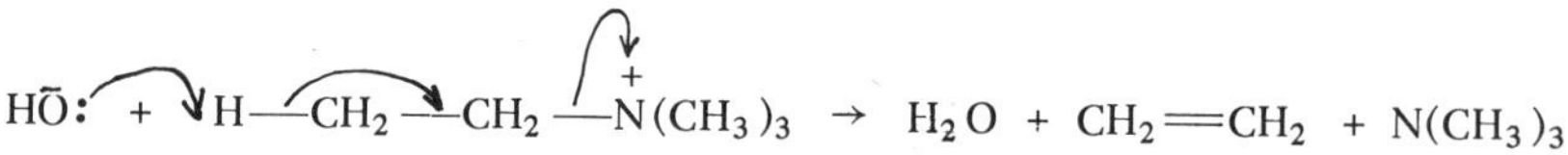

4. Debromination

Alkenes are also produced by the reaction of zinc dust with an alcoholic solution of a 1,2-dibromoalkane. For example, ethene is formed by the action of the metal on 1,2-dibromoethane:

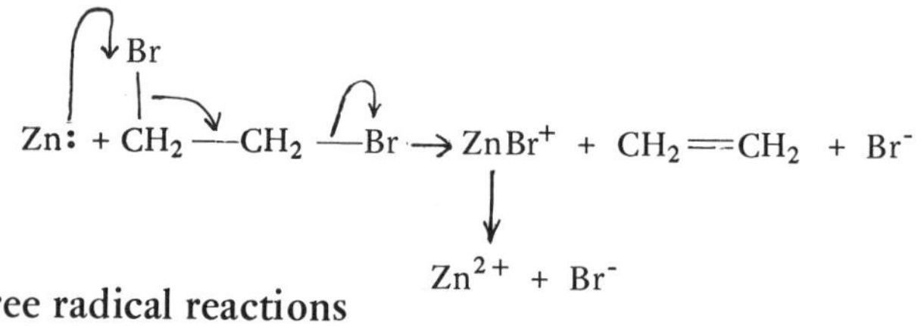

Free radical reactions

Free radicals can be produced under a number of conditions – at high temperatures (in a flame), by ultraviolet radiation and by the use of peroxide initiators.

Ultraviolet radiation breaks the bond in chlorine molecules:

$$Cl_2 \xrightarrow{\text{u.v.}} 2\,Cl\cdot \qquad \Delta H = 242 \text{ kJ mol}^{-1}$$

These atoms, or radicals, each have an unpaired electron and react readily with each other (to reform chlorine gas) or with hydrocarbons to abstract hydrogen atoms:

$$Cl \cdot + CH_4 \rightarrow HCl + CH_3 \cdot \quad \Delta H = -4 \text{ kJ mol}^{-1}$$

The methyl radicals can combine with other radicals or they can attack more chlorine molecules:

$$CH_3 \cdot + Cl_2 \rightarrow CH_3Cl + Cl \cdot \quad \Delta H = -101 \text{ kJ mol}^{-1}$$

Peroxides are used to initiate reactions such as polymerisations. In these, the peroxide group fragments to give oxygen radicals which act in a similar manner to that described in the chlorination reactions:

$$R\text{---}O\text{---}O\text{---}R \rightarrow RO \cdot + \cdot OR$$

Summary

At the conclusion of this chapter, you should be able to:

1. account for the polarity of covalent bonds;
2. distinguish between the +I and −I effects;
3. relate the relative strengths of organic acids to the inductive forces present in the molecules;
4. account for the relative strengths of bases by reference to the inductive forces present;
5. explain why acyl halides are more reactive than halogenoalkanes;
6. state that, due to the delocalisation of the π-electrons, the carboxylate ion is symmetrical;
7. classify reagents as electrophilic, nucleophilic or free radicals;
8. describe the differences between S_N1 and S_N2 reactions, stating the conditions that favour each;
9. describe the mechanism of the electrophilic substitution of benzene by polar and non-polar reagents;
10. describe the mechanism for the electrophilic addition of a molecule to an alkene;
11. describe the mechanism of the nucleophilic additions to the carbonyl group;
12. state that condensation reactions are addition−elimination reactions in terms of their mechanism;
13. describe the mechanism of elimination reactions yielding alkenes;
14. specify the conditions under which an elimination reaction is preferred to a substitution reaction.

Experiments

8.1 Mechanism of halogen addition to alkenic bonds

If the mechanism for addition is correct (p. 231), then after the initial addition of the bromine cation

$$Br^+ + \begin{array}{c} CH_2 \\ \| \\ CH_2 \end{array} \rightarrow \begin{array}{c} CH_2 \\ {}^+Br\diagup \big| \\ \diagdown CH_2 \end{array}$$

the bromonium ion will rapidly react with any negatively charged ions available. Normally this would be Br⁻, but water is also a nucleophile,

$$H_2O: + \begin{array}{c} CH_2 \\ | \quad Br^+ \\ CH_2 \end{array} \xrightarrow[-H^+]{} \begin{array}{c} HO-CH_2 \\ | \\ CH_2-Br \end{array}$$

generating a bromohydrin. In the presence of NO_3^-, a nitrate may also be formed:

$$NO_3^- + \begin{array}{c} CH_2 \\ | \quad Br^+ \\ CH_2 \end{array} \rightarrow \begin{array}{c} O_2N-O-CH_2 \\ | \\ CH_2-Br \end{array}$$

The nitrate must be present before the addition of the bromine as the second stage is rapid.

To a solution of pent-1-ene (1 cm³) in ethanol (15 cm³), add a solution of sodium nitrate (3 g) in water (10 cm³). Shake to produce a homogeneous solution and then add bromine (0.35 cm³) – carefully. Dilute with water (20 cm³) and allow the organic layer to settle. Decant off the aqueous layer and extract the organic products into ethoxyethane. Dry over magnesium sulphate and evaporate off the solvent in a warm water bath. Analyse the oil for bromine and nitrogen in the mixed products.

References

Organic Reactions; J. G. Stark (Pergamon, 1970).
A Guide to Understanding Basic Organic Reactions; R. C. Whitfield (Long-man, 1966).
Guidebook to Mechanism in Organic Chemistry; P. Sykes (Longman, 4th edn. 1975).
Evidence for some organic reaction mechanisms; M. A. Atherton and J. K. Lawrence, *Sch. Sci. Rev.,* 1969, **51**, 295–308.
Nucleophilic substitution – a student experiment; A. Yeadon; *Educ. Chem.,* 1970, **7**, 241–2.

Films

Mechanism of an organic reaction (CHEM Study).
Molecular models and substitution reactions (Centre for Educational Practice, University of Strathclyde).

Questions

1. Account for the following observations:

   (a) ethanoyl chloride is much more reactive than chloroethane;

   (b) 2-chloroethanoic acid is stronger than ethanoic acid;

   (c) 3-aminopropanal is significantly different in its chemistry from propanamide;

   (d) aminoethane has a pK_b value of 3.07; ethanamide has a pK_b value of 15.1;

   (e) aniline (aminobenzene) is a stronger base than 4-nitroaniline.

2. Suggest a mechanism for each of the following changes:

   (a) $CH_3 CH_2 I \xrightarrow{OH^-} CH_3 CH_2 OH$;

   (b) $(CH_3)_3 CBr \xrightarrow{OH^-} (CH_3)_2 C{=}CH_2$;

   (c) $(CH_3)_2 C{=}CH_2 \xrightarrow{HBr} (CH_3)_3 CBr$;

   (d) $CH_3 CH{=}CH_2 \xrightarrow{Br_2} CH_3 CHBrCH_2 Br$

3. Describe the mechanism for the addition of the following reagents to a carbonyl compound:

   (a) $CH_3 MgBr$;

   (b) HCN;

   (c) $Na^+HSO_3^-(aq)$

4. Describe the mechanism of the following series of reactions:

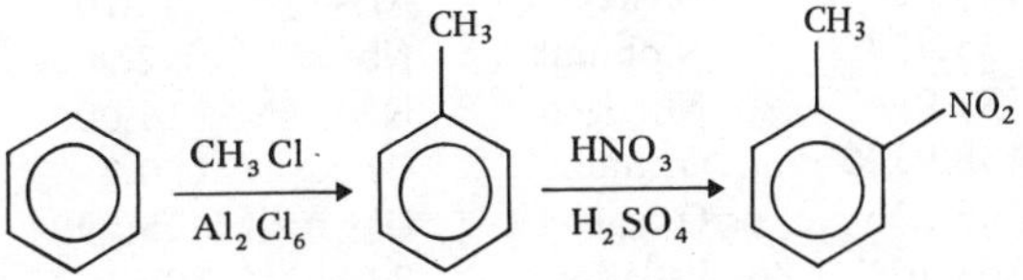

Appendix

Relative atomic masses

Element	Symbol	R.A.M.	Element	Symbol	R.A.M.
Aluminium	Al	26.98154	Iridium	Ir	192.22
Antimony	Sb	121.75	Iron	Fe	55.847
Argon	Ar	39.948	Krypton	Kr	83.80
Arsenic	As	74.9216	Lanthanum	La	138.9055
Barium	Ba	137.34	Lead	Pb	207.2
Beryllium	Be	9.01218	Lithium	Li	6.941
Bismuth	Bi	208.9804	Lutetium	Lu	174.97
Boron	B	10.81	Magnesium	Mg	24.305
Bromine	Br	79.904	Manganese	Mn	54.9380
Cadmium	Cd	112.40	Mercury	Hg	200.59
Caesium	Cs	132.9054	Molybdenum	Mo	95.94
Calcium	Ca	40.08	Neodymium	Nd	144.24
Carbon	C	12.011	Neon	Ne	20.179
Cerium	Ce	140.12	Nickel	Ni	58.70
Chlorine	Cl	35.453	Niobium	Nb	92.9064
Chromium	Cr	51.996	Nitrogen	N	14.0067
Cobalt	Co	58.9332	Osmium	Os	190.2
Copper	Cu	63.546	Oxygen	O	15.9994
Dysprosium	Dy	162.50	Palladium	Pd	106.4
Erbium	Er	167.26	Phosphorus	P	30.97376
Europium	Eu	151.96	Platinum	Pt	195.09
Fluorine	F	18.99840	Potassium	K	39.098
Gadolinium	Gd	157.25	Praseo-		
Gallium	Ga	69.72	dymium	Pr	140.9077
Germanium	Ge	72.59	Rhenium	Re	186.207
Gold	Au	196.9665	Rhodium	Rh	102.9055
Hafnium	Hf	178.49	Rubidium	Rb	85.4678
Helium	He	4.00260	Ruthenium	Ru	101.07
Holmium	Ho	164.9304	Samarium	Sm	150.4
Hydrogen	H	1.0079	Scandium	Sc	44.9559
Indium	In	114.82	Selenium	Se	78.96
Iodine	I	126.9045	Silicon	Si	28.086

Element	Symbol	R.A.M.	Element	Symbol	R.A.M.
Silver	Ag	107.868	Tin	Sn	118.69
Sodium	Na	22.98977	Titanium	Ti	47.90
Strontium	Sr	87.62	Tungsten	W	183.85
Sulphur	S	32.06	Uranium	U	238.029
Tantalum	Ta	180.9479	Vanadium	V	50.9414
Tellurium	Te	127.60	Xenon	Xe	131.30
Terbium	Tb	158.9254	Ytterbium	Yb	173.04
Thallium	Tl	204.37	Yttrium	Y	88.9059
Thorium	Th	232.0381	Zinc	Zn	65.38
Thulium	Tm	168.9342	Zirconium	Zr	91.22

Answers to numerical questions

Answers

Chapter 1 (p. 28)

4. 55.8660   6. -159 kJ mol^{-1}

Chapter 2 (p. 51)

2. 0.045   3. 3.38   4. 10.2   10. (a) -212 kJ mol^{-1}; (b) 9.04×10^{36};
(c) $+36.3$ J deg^{-1} mol^{-1}

Chapter 3 (p. 84)

3. 1   5. 143.6 kJ mol^{-1}

Chapter 4 (p. 122)

3. 0.15 g

Chapter 5 (p. 151)

1. (a) 0.104; (b) 0.070; (c) 2.80   2. 99.2 per cent   3. 6   5. 36.1 per cent.

Index